Studies in History and Philosophy of Science
Volume 65

Studies in History and Philosophy of Science is a peer-reviewed book series, dedicated to the history of science and historically informed philosophy of science. The series publishes original scholarship in various related areas, including new directions in epistemology and the history of knowledge within global and colonial contexts. It includes monographs, edited collections, and translations of primary sources in the English language. These cover a broad temporal spectrum, from antiquity to modernity, and all regions of the world.

Juan A. Queijo Olano • Evelyn Mozo Meneses

Giants Upon Our Shoulders: A Philosophical History of Physics in Uruguay

Springer

Juan A. Queijo Olano
Department of History and Philosophy of Science
Universidad de la República
Montevideo, Uruguay

Evelyn Mozo Meneses
Department of Philosophy
National Administration of Public Education
Montevideo, Uruguay

This work was supported by Institute of Physics, Faculty of Engineering, University of the Republic.

ISSN 1871-7381 ISSN 2215-1958 (electronic)
Studies in History and Philosophy of Science
ISBN 978-3-031-97728-2 ISBN 978-3-031-97729-9 (eBook)
https://doi.org/10.1007/978-3-031-97729-9

This Springer imprint is published by the registered company Springer Nature Switzerland AG.
The registered company address is: Gewerbestrasse 11, 6330 Cham, Switzerland

Preface

This work arises from the collaboration between the Institute of Physics of the Faculty of Engineering, and the Department of History and Philosophy of Science of the Faculty of Humanities and Educational Sciences, both of the University of the Republic. This collaboration has allowed the funding of this research to build a history of physics in Uruguay. Like any history, it is incomplete and deserves future revisions, improvements, and complements, and therein lies the ultimate purpose of this work: to open up the possibility of further work to generate a history of science in a country that has historically provided little room for this dimension of human intellectual activity.

The other main purpose is to promote a more assiduous and lasting work between physics and philosophy. The physics community in Uruguay has been achieving, not without many obstacles and difficulties, a development that has been internationally recognized, especially in areas such as quantum field theory and general relativity. These developments, innovative from the theoretical and conceptual point of view, are necessary case studies for a philosophy of science that is concerned with understanding scientific developments in their contexts and times. Philosophy of science has a lot of work ahead of it if it is to be a genuine actor contributing to the better scientific development of countries historically far from state-of-the-art development.

With much modesty, we hope that this work will be understood as one of the many contributions that Uruguayan and Latin American science needs, in order to become a dimension of social and collective life that allows us to build fairer societies full of opportunities for all, especially for the most excluded.

Montevideo, Uruguay — Juan A. Queijo Olano

Acknowledgments

This book is the product of two years of deep research on the history of physics in Uruguay. This work began three years ago from the joint interests of the Department of History and Philosophy of Science of the Faculty of Humanities and Sciences of Education (Udelar) and the Institute of Physics of the Faculty of Engineering (Udelar), and concludes with these pages. We want to express deep gratitude to Nicolás Wschebor, Italo Bove and Alejandro Romanelli—from the Institute of Physics—for the trust placed and the constant support throughout this time. We also thank our colleagues from the Institute of Philosophy: Ma Laura Martínez, Matías Osta, León Trelles, Alejandro Chmiel, Álvaro Peláez and Guillermo Nigro.

Throughout this work, we had the support, corrections, suggestions, and discussions proposed by Professor Antonio Augusto Passos Videira, from the State University of Rio de Janeiro and the CBPF. Through his generosity and his enormous knowledge of Latin American physics, we have been able to carry out, we believe, a better job than we would have done without his presence.

Many people and teams provided us with invaluable support during the conduct of this research: the team from the General Archive of the University of the Republic, for their continuous availability to provide access to historical resources indispensable for this research. To Mariangela Santurio, a member of the team from the Historical Archive of the Faculty of Medicine, for helping us in the search for essential documents to understand the relationships between physics and medicine in Uruguay. To Marcelo Marmoria, in charge of the Central Archive of the Faculty of Engineering, for his willingness to answer queries and guide us in accessing the documentary collection of physics in the Faculty of Engineering. Also we thank Mónica Pagola, Gonzalo Marín and Pablo Darriulat, members of the Central University Archive of the Faculty of Humanities and Sciences of Education.

Acknowledgments from Evelyn Mozo Meneses:

To the women of my family, my grandmother Nibya, my mother Nancy, my sisters Maira and Ma. Nancy, and my nieces Julieta and Uma—thank you for your unconditional support, your infinite affection and for being my company and support at every step of this academic and personal journey.

To my friends Alexis, Bruno, Lucía Á, Lucía S, Pía and Valentina, who were always there to listen to my ideas, share my doubts and celebrate my dreams. Your friendship gives a special meaning to everything I do.

To Juan A. Queijo, with whom I had the privilege of making this book. In addition to being a great workmate, he is also the main engine of this project. Thank you for your trust, dedication, and inspiration, for the shared ideas and for making this journey an enriching experience.

And to the teachers and colleagues who were part of my professional training, for their guidance, teachings, and the example they left at every step of my journey. My contribution in this work is also a reflection of what I learned from you.

Acknowledgments from Juan A. Queijo Olano:

My first thanks is to Evelyn, my companion on this journey of research. Her work and philosophical dedication, I am sure, has allowed me to be a better researcher than I was at the beginning.

Getting involved in this type of academic adventures, I understand, is only possible if a large support network allows us to dedicate days and nights to research. I would like to thank my parents, Daniel and Gloria, for their constant unconditional support. To Cecilia and her sudden and bright entrance. Finally, I dedicate this enormous effort to my dear daughter Franka, a source of immeasurable energy.

Contents

Abbreviations

AFA	Argentine Physics Association
AUPCC	Uruguayan Association for the Advancement of Science
CBPF	Brazilian Center for Physical Research
CLAF	Latin American Physics Center
CNEA	National Atomic Energy Commission
CONICYT	National Council for Scientific and Technical Research
FHyC	Faculty of Humanities and Sciences
FIA	Faculty of Engineering and Surveying
IES	Institute of Higher Studies
IPA	Artigas Teachers Institute
U.T.E	Administration of State Power Plants and Telephones
UBA	University of Buenos Aires
Udelar	University of the Republic (Uruguay)
UNESCO	United Nations Educational, Scientific and Cultural Organization
UNLP	National University of La Plata (Argentina)
UNT	National University of Tucumán (Argentina)
USA	United States
USP	University of São Paulo (Brazil)

Introduction

The phrase that serves as a reference in our title is often attributed to Isaac Newton. Within his letter to the English natural philosopher Robert Hook, a phrase will remain for posterity in the imagination of modern science: "If I have seen further, it is by standing on the shoulders of giants." The image evoked by this phrase corresponds to a tradition of science and the history of science that not only constituted what we could call, following Robert. K. Merton, the "ethos" of science; it was also the image that Western societies granted to scientific activity to give it a specific cultural meaning. Thus, the image we have built of science speaks to us of an activity that is erected on the basis of a universal, extemporaneous, and open cooperation; a cumulative activity where individual interest has to correspond with that of this "giant" -science-; of an activity that allows us to see "further" only to the extent that we stand on what is accumulated and are willing to offer what we see for the purpose of continuing to grow the "giant".

We could continue indefinitely with the metaphors that the phrase used by Newton enables us, but the idea, in general, is known to all. And it is also because, at least since the middle of the twentieth century (if not before), that the *giant* image with which science was defined began to be the object of criticism. This critical movement toward the definitions of what science was and is, the discernment of what constitutes its center and specificity, coincided with the opening and expansion of areas that were interested in this type of debate. If the European and modern image of science evoked by Newton was used, assumed, and disseminated by the scientific community itself over these last centuries, to the critical movement of that image we can attribute in addition to its inquisitive intervention, the snatching of the authority that granted legitimacy exclusively to scientists to speak on behalf of science.

We are not interested, in this introduction, in making a precise and exhaustive chronicle about the history of the philosophy of science, its debates, and its main exponents. On the one hand, many of these narratives have already been constructed and proposing another would imply something that is outside our interest.[1] But on the other hand, we consider that the development of the philosophy of science cannot be explained, or it is pointless to do so, in exclusively philosophical terms. By proposing a philosophical history of physics in Uruguay, we want to place this work within the framework of the naturalistic approaches of the philosophy of science. In other words, we understand that reflection on science cannot be detached and treated in the same way as philosophical concepts are worked with. Science is not a philosophical concept, it is the summarized expression of a very varied set of activities that manifest in very different ways. Only a philosophy that approaches these activities will be able to account for the philosophical problems they contain; therefore, more than a general philosophy of science, we understand that the path of the philosophy of science must be traced by the particular approaches of disciplines such as the philosophy of physics, philosophy of chemistry, mathematics, or biology, among others. It is clear that a naturalized philosophy of particular sciences has no justificatory pretension, but above all a comprehensive one. This issue is central in the history of physics that we present in this book: it is not about justifying under universal criteria the scientific character of what the development of physics in Uruguay and other parts of Latin America implied. On the contrary, by wanting to approach the practices of physics in these regions we seek to better understand how they were carried out by their actors, the institutions, and the culture that supported them.

From the middle of the twentieth century, between Europe and the United States, a way of thinking about science emerged that problematized many aspects related to what scientists do: how they truly explain their phenomena, how they construct their reasoning, where their authority and social legitimacy are founded, etc. These reflective exercises marked an era of debates, many of them very hard, from which we believe that science that had never been considered by anyone benefited. By problematizing those foundations of the construction of scientific knowledge, the edifice of science was also questioned. A consequence of these tremors was that, perhaps, the idea of science that the official history of science had constructed had dedicated itself more to justifying than to understanding.

Among these tensions, the look at scientific practices that had historically been relegated was born. A science that far from being able to reach the shoulders of the giant, always carried its historical, model, and enchanted weight. This work is an exercise to tell the story of the developments of *dwarf* science, but with the explicit purpose that, as the reader delves into it, they can observe it for its own merits and defects, limitations, and achievements.

---

[1] Just to mention some recent works: Ladyman 2002; Godfrey-Smith 2009; De Witt 2010; Kosso 1998.

What is Physics?

That science that we have studied is the *physics* that developed from the nineteenth century in Uruguay. But what exactly is physics? This is not a rhetorical, empty question. In his book *The Philosophy of Physical Science*, Sir Arthur Eddington recognizes the problem of being able to constrain a definition of this discipline in a taxative way:

In principle, we might identify physical knowledge with the contents of certain encyclopedic works, such as the *Handbuch der Physik*, which between them cover the various branches of physical science. But there are obvious objections to a slavish acceptance of a particular authority; and I will therefore define physical knowledge to be that which a right-thinking person would today accept as justified by physical science. [2]

The first reason that led us to this study is the relative general ignorance that has existed about how science developed in Uruguay, and in particular physics. It was the interest of a new generation of young physicists and philosophers of science in showing a history completely absent from the cultural and intellectual production of Uruguay.

The second reason is to better understand Latin American science, knowing it in its details, observing its particularities, and taking, as a case study, not those universally known successful episodes, but those trajectories that did not involve giant heroes of knowledge, or great results or discoveries. More modestly, we are interested in thinking, based on the existing documentary evidence, that a definition of science can emerge from knowledge about how beyond the cultural centers of the world purposes were also pursued to establish a knowledge that serves for the construction of the most relegated nations. Physics in Uruguay was that science that was destined to survive, to sustain itself based on individual impulses, to build its own networks and scientific communities from the resources it had at its disposal. Physics was resisted for many years, little appreciated by the predominantly literary and humanistic intelligentsia, those who did not understand the channels of the science of their time, nor its cultural relevance. But even in this context, which marks the national identity of Uruguay so far from science, there were episodes where some actors and certain policies allowed to erect scientific developments that were the beginnings of current Uruguayan science.

Seeking to know and understand physics in Uruguay has not been a solitary enterprise. The first reason is that, as we will point out throughout the book and in the conclusions, there is no *strictu sensu* a Uruguayan physics if we do not seriously question the very category of " science", just as there is no science delimited by the circumstances of political geography. As paradoxical as it may be, the unique science of Latin American countries was also a global science, but with a sense and content that are its own. Therefore, this historical journey is never exclusively

[2] (Eddington 1939, 2).

Uruguayan. It is a journey that involves the entire southern region of America, and even all of the Americas. In this, we have benefited from the work of fellow historians and philosophers of science who have been contributing to a better understanding of Latin American science by knowing its history. There is also the benefit of having had the support of current physicists from various parts of the continent, always willing to dialogue and inform us about their knowledge. They have recognized that little is known about the past of this discipline to which they dedicate so much time, and they have encouraged and supported us to write this history.

The book consists of six chapters and the density of each one of them is closely related to the documentary material that supports what we have been able to know about each period and each area of physics. In any case, as was mentioned, we have been forced to break the geographical borders on many occasions, because the development of physics depended less on institutional impulses than on the life trajectories of its protagonists.

Chapter 1 focuses on the dawn of the Eastern Republic of Uruguay, more precisely on the foundation—in 1849—of what has been, since the beginning of the nation's independence, its main house of studies: the University of the Republic (Udelar). In some sense, this work follows the trajectory of physics at the Udelar, but it also surpasses it many other times. In these beginnings, we find that, among the limited knowledge that was offered, there was training in *general physics* and *particular physics*, a direct heritage of a philosophical branch of the River Plate region called "physicism". In this context, physics is a malleable knowledge, adapted to various practices and ideas, and although it managed to establish itself as a legitimate knowledge within university studies, its development and relevance were still not well defined.

The second chapter shows the impact of two institutional transformations in the education of physics, and some beginnings of research. In 1875 and 1888, respectively, the Faculties of Medicine and Mathematics and Related Branches were founded, which shows that the inclusion of physics responded to a growing appreciation of the sciences in the construction of a national project. In both careers physics knowledge is included, oriented to medicine, in the first, and to industry, in the second. Even from a perspective slightly inclined towards the problems that those decades presented in terms of understanding certain natural phenomena, some professors try to learn from new discoveries and encourage certain practices in university classrooms.

The turn of the century meant a real break in Uruguayan science, and especially in physics and mathematics. In the recently created Faculty of Engineering, a split from that first Faculty of Mathematics, a group of professors emerges who consider it necessary to reform the mathematics that are taught in the country. From this simple change of curricular plans, a new type of engineer capable of understanding the new challenges of physics and its mathematization is derived. In this incipient context, Einstein's arrival in the country meant a fundamental point for physics, until then rejected in the environment of engineers for being a training of "luxury" for the material conditions of the country. Simultaneously, the foundation of the Institute of Radiology at the Faculty of Medicine represented a significant contribution to experimental physics, highlighting its practical application and its contribution to medical advancement in the country. From there, basic science began to consolidate in Uruguay, oriented toward the deep study of nature and its phenomena. This is the central focus of the third chapter.

In the fourth chapter, we focus on the set of protagonists who directly or indirectly tried to engage with the new theory of relativity. In one of its most important central squares in Montevideo, Uruguay preserves a famous statue where the meeting of Albert Einstein with the Uruguayan philosopher Carlos Vaz Ferreira in 1925 is recreated. This monument represents the meeting of international scientific genius with the most relevant intellectual personality of the country. But it shows how little the country was aware of what the new theories of modern physics proposed. In this section, we seek to show a deeper image of the cultural impact of relativity in Uruguay, highlighting the group of engineer-physicists who understood and worked on the theory.

The fifth chapter stops and explores the diversity of practices that had physics at their center, their main promoters, and the scientific links they generated, in the first half of the twentieth century. We can observe a plurality of physics experiences, such as the attempt to build research in theoretical and nuclear physics, integration into cosmic ray networks, innovative experiences in nuclear medicine and spectrography, and new actors of international relevance in the fields of relativity and field theory, or astronomy. This chapter covers all these impulses, their main protagonists, and the achievements reached for a country still reluctant to think of science in a more structural way and connected with national development. But this era also shows a substantial change, a product of the end of the war and the global threat that military research in nuclear energy represented. Some of these physicists were part of public debates, integrated philosophical positions in defense of science (learned in their contacts with philosophers and scientists from abroad), and condemned the weight that the literary and humanist tradition still had in the country.

Finally, we reach the final stage of the history of this book, which covers the 1960s until the arrival of the dictatorship in the country. This military process that erupted in 1973 cost the scientific and academic communities of Udelar a great deal. The political persecution of some professors led to the exile of many of them, and the reconstruction of the scientific networks that existed during this period until the democratic recovery, in 1985, are not considered in this book. But we can observe that, before the Coup d'Etat, especially in the 1960s, the discussion and projects for the development of basic sciences were imposed at the University. This discussion was not only national. Particularly in physics, it involved actors from the region who participated in the discussions about what model of science should be installed in Uruguay. It was a plan for the development of basic sciences that, had it not been for the military intervention in the country and the university, would have meant a distinctive milestone for Uruguayan science.

The final chapter leaves the historical work to focus on some philosophical lines on how to conceive, understand, and think about science from countries historically isolated from international scientific developments, and from the great narrative of the history of science. In a way, the historical research that precedes the conclusions may be just the excuse (a necessary, indispensable excuse) for what interests us: to contribute to understanding the current difficulties of science in Latin America from a serious study of its history. Throughout the years in which we have developed this work, a maxim of Goethe, timidly appearing in his *Sketch of a Theory of Colors* has guided us: "the history of science is science itself".

Chapter 1
Genealogies of a Discipline in the Río de la Plata

To reconstruct the History of Physics in Uruguay, it is unavoidable to take as a reference the creation of the University of the Republic (Udelar), in 1849. This is important because this creation occurs almost parallel to the national construction of the Uruguayan state, in 1830. This leads the historical studies of science and knowledge in Uruguay to show the presence of its only university (this was the case until the end of the twentieth century) as the main space where the disciplinary development of the sciences took place. In the nineteenth century, certain attempts to insert a discipline such as physics—still not very autonomous in the world—constantly failed in the dispute to place the natural sciences at the very heart of the educational institution. This presents a double difficulty for this type of research: on the one hand, the difficult acceptance of the role that the natural sciences could play in the construction of a national ruling elite, since the first purpose of the Universidad Mayor de Montevideo was the formation of a ruling class for the nascent state (mainly this formation was thought in jurisprudence, which was the branch cultivated par excellence in the first century of existence of the University); and on the other hand, it becomes problematic to recognize the isolated attempts of certain actors who sought to convince and carry out some concrete experiences of scientific training.

The Uruguayan philosopher Arturo Ardao has identified five manifestations of philosophical thought in Uruguay, which precede the creation of the Udelar: scholasticism, rationalism, the encyclopedia, ideology, and Saint-Simonianism.[1] He suggests that these correspond to various stages of the nation's formation, although they do not necessarily exclude each other and may overlap in some periods. Ardao's work tells us something very significant, which deserves our attention when we want to unravel the most ancient scientific practices in our territory: his work

[1] Ardao (1945).

J. A. Queijo Olano, E. Mozo Meneses, *Giants Upon Our Shoulders: A Philosophical History of Physics in Uruguay*, Studies in History and Philosophy of Science 65, https://doi.org/10.1007/978-3-031-97729-9_1

focuses the analysis on the characters of an intellectual elite—either European or trained in the old continent—that shows almost no interest in a philosophy of nature. In other words, the characters and thoughts that occupy Ardao's work, and which are the intellectuals who will forge the destinies of a Universidad Mayor de Montevideo, belong to a learned tradition where thought about nature is not of central attention. But even so, we can see, with documentary clarity, that studies on physical phenomena occupied a place within the study programs of the various educational spaces in Montevideo. First, they were part of the philosophical training that was taught in schools, it was then a solitary chair for a long time when university studies were instituted, and it was finally integrated into the first Faculties of Medicine and Mathematics created in the last quarter of the twentieth century, showing the applicable benefits of their studies. This allows us to recognize, at least tentatively, that there is a continuity of what was understood by *Physics*—a conceptual unit not always clearly defined, but symbolically relevant within the set of knowledge to which university training should aspire. This continuity will be subject to the ways in which this knowledge sought to meet various concrete needs at different stages of the institutional construction of the University. In other words, if there is a possibility of reconstructing a history of Physics in the Udelar, we must assume that with that term different areas of knowledge were signified, not always clearly delimited.

1.1 The first Rioplatense Plot of Physics: Physicists

We can recognize, in the pre-university stage of Montevideo, the names of José Benito Lamas (1787–1857), the Argentine Alejo Villegas (1783–1857) and the Uruguayan naturalist priest Dámaso Antonio Larrañaga (1771–1848) as those first precursors of physics studies in the country, based on the training they were able to receive. What unites these characters is having been able to participate in a particular type of philosophical development—which must be placed within the cultural framework of Rioplatense exchange that takes place towards the end of the eighteenth century[2] and which was called *Physicism*. Although circumscribed within philosophical teaching, this perspective recognizes physics as autonomous in the intellectual field. Following this tradition, little recognized in the literature of intellectual history and ideas, implies adhering more firmly to the existence of communities and characters than to the canon of institutions. This type of work in the history of science is recognized under the label of contemporary *Historical Epistemology*, which maintains its links with the types of historiography mentioned, but maintaining the peculiarity of emphasizing the production derived from the practice of science—books, notes, letters, study programs, etc—carried out by

[2] De Asúa (2010).

scientists.[3] Let's start to weave that plot. The manuscripts of the priest José Benito Lamas, according to Ardao "the most characteristic representative of scholastic teaching in the country", tell us about an encyclopedic training that covered studies on arithmetic, nautical, philosophy, theology, and metaphysics. During his stay in Montevideo in the first decades of the nineteenth century, he consecrated a philosophy course that dealt with Metaphysics and Logic (in its first year, 1833) and about Ethics or Moral Philosophy and Physics (in the second year, 1834). There were eight students involved in that cohort. The way The evaluation followed the precepts of the scholastic tradition: the students were summoned on a certain date, where they were assigned the presentation they had to give (based on a series of already stipulated philosophical premises), and those dissertations that they had to discuss with their other classmates. All of this usually happened in an open-door setting and the panel was generally made up of five distinguished professors. The president of the public event on that occasion, Don Joaquín Campana, described in his summary what were then the studies of physics within a philosophical education:

> And in Physics, which embraces the general theory of matter in which its divisibility, impenetrability and porosity are observed in the light of experience and discourse. The general theory of movement in which everything concerning the estimation of movement, its obstacles, its general laws and its communication, both in elastic bodies and in those that are not, is examined and explained successively. The theory of some particular qualities of matter such as cold, heat, taste, sound, light and colors, and the Theory of the Sky in which the systems of Ptolemy and Tycho Brahe are challenged for being contrary to astronomical observations, the one of Copernicus is admitted, as more in accordance with them.[4]

One of the members of the student evaluation committee was Dr. Alejo Villegas, who had a philosophical education at the San Carlos College in Buenos Aires, earning his doctorate in 1804 and 1806 in Theology and Civil Law, respectively. We will return to him later, to account for the Physics programs he taught on the occasion of the Philosophy courses at the House of General Studies in Montevideo, in 1839 and 1840. At the public event held in 1834, Villegas-as evaluator-made a tour of the process of training in Philosophy that these students had gone through:

> We already have a group of young people, who have so profitably completed the course of their preparatory studies, skilled in the art of seeking truth by the rules of good Logic: persuaded in the existence of other immortal beings, which are not within the sphere of our bodily senses, according to the knowledge that metaphysics has provided them; educated in the elementary principles of natural synderesis by the study of Ethics, to guide their reason on the path of the healthiest morality: and finally convinced by Physics of matter and bodies, which from heaven to earth, and everywhere, surround our existence, beautifying the universe, and filling human intelligence with admiration, both for their particular attributes and movements, and for the coordination of one with the other. Such is the capital of lights and knowledge acquired in this philosophical course.[5]

---

[3] Ardao (1945, 38).

[4] Lamas (1836, 29–30).

[5] Lamas (1836, 34).

Physics encompassed a part integrated into the analysis of matter, the study of movement and the astronomical knowledge acquired from the Copernican revolution. But how did these topics become part of the philosophical education of Montevideo's youth? Montevideo, towards the end of the eighteenth century, had a population of about 4470 inhabitants and it was estimated that Buenos Aires quintupled it in number.[6] These demographic figures will only grow throughout the nineteenth century, mainly motivated by the urban organization of the new republics and the formal establishment of ports, arrival of large influxes of immigrants. In this sense, by 1830, with Uruguay's independence already assumed, many immigrants arrived, mainly from Europe and the Argentine provinces. Given this general situation, we can infer that the training in Physics inscribed in the first formations in Philosophy in our country have a direct link with the movement *physicist* of the San Carlos College of Buenos Aires. In fact, this Buenos Aires college acquires a notable relevance to understand the first characters who approached the problems of physics and natural science in Montevideo. The analysis of these links becomes central in the sense that the Carolina training in Buenos Aires meant a marked attempt to break with all scholastic philosophy still hegemonic in universities in America. And it is particularly the studies of General Physics and Experimental Physics that were the heart of the fight against this scholastic tradition. Some writings of the Buenos Aires intelligentsia were very critical against the scholastic expression that was taught in that House of Studies. For example, the professor of Philosophy at the College between 1797 and 1799, Manuel Gregorio Álvarez, denounced the attachment to Aristotelianism still reigning in the teaching of philosophy in that establishment:

> If we ask Aristotle what is the nature of a diaphanous body, he will answer without hesitation and in a masterful tone, that it is that by which the body constitutes the species of diaphanous, and if we ask him again what is that which constitutes the body in the species of diaphanous, he will tell us that it is diaphaneity. Let's ask him again, since he has not exhausted the wealth of his answers, why a being exists, and he will tell us that it is because of its esseity; why it is an animal, and he will tell us that it is because of its animality, etc. His science does not go beyond that. It is certainly surprising, and today we marvel at it, that after a thousand and more years, this sect [Peripatetic] still reigns in schools, tyrannizing understandings, which its followers no longer know anything about forms other than that the form of stones is petreity, that of iron is Ironness, the one of cheese, cheesiness.[7]

In the same tone a few years later, Dr. Mariano Moreno, an Argentine politician and journalist also educated at the San Carlos school, denounced the state of teaching, which was not up-to-date with the findings of the science of his time, hinting at the imprint of changes that were glimpsed in the education of those who passed through said institution:

> As for the utility that was to be expected from promoting knowledge and sciences, with their lessons reduced to forming of the students intolerable theologians, who spend their time in stirring and defending abstract questions about divinity, angels, etc, and consume

[6] Duffau (2016).

[7] Furlong (1947, 345).

> their lives in finding out the opinions of ancient authors who have established extravagant and arbitrary systems on points that no one is capable of knowing, we must say that it is absolutely none.This principle of misguidance of ideas for the studious youth, could be compensated by the advantages of instructing in the branches of logic, natural and experimental physics, ethics and metaphysics that are taught to students for a period of three years, before moving on to Theology, which as the most necessary and what they should get freshest from their heads, is left for last. But it is painful to add that in these branches scholasticism is still observed in all its rigor, and that even theses that have been abandoned in Europe 50 years ago are still defended, or the discoveries made by moderns are ignored, in this very profitable part, of human knowledge. I have seen professors who could speak with the greatest propriety about any physical matter, be entirely embarrassed in the presence of a pneumatic machine, or the apparatus for the dissolution of gases, whose theory they could explain admirably, but in no way execute.[8]

Training in physics generally gathered a series of maxims, or "truths", which were the summary of both those most ancient knowledge as well as the most recent novelties of the time. We can anticipate that, contrary to what is usually signified, the knowledge of these *physicists* was quite up to date with the significant changes that were recorded in Europe in those centuries. On both banks of the Río de la Plata, in relation to forming communities dedicated to the new philosophy, we can recognize a body of students and teachers who-not without difficulties-modestly establish themselves as a community, by giving value to that branch of philosophy closest to natural philosophy.

1.2 Institutionalisation (1790–1850)

Two stages of physics, separated by a few decades and by the Río de la Plata, make up the academic beginning of this discipline. On one hand, in a study conducted by historian Celina Ana Lértora Mendoza on the *General Physics* taught by Dr. Estanislao de Zavaleta, at the San Carlos College, between 1795 and 1797. On the other, the courses of General and Particular Physics by Alejo Villegas, at the House of General Studies in Montevideo, in the years 1839 and 1841. Celina Lértora records Zavaleta's work as that of a scholastic who, in a not very dogmatic way, includes the innovations of the scientific advances of his time. His physics is presented in close relation with metaphysics, in the sense that the former "is required as crowning and culmination" of the latter.

> (...) he does not come to accept a sharp separation between science and philosophy, and the strictly scientific problems of physics will be incorporated as appendices to the classic themes of cosmology. In all this Zavaleta is extremely coherent, as we must recognize that both in physics and in metaphysics, the incorporation of modern themes was done with great care, without breaking the fundamental unity of thought, but also without altering these theories and taking them out of their true sense and context.[9]

---

[8] Moreno (2004, 15).

[9] Lértora (1972, 7–40).

The structure of the General Physics program adopts the classic form of scholastic teaching, as was mentioned. To the presentation of a problem, follow the journey through the opposing positions that have existed as a response, and a conclusion—which is presented by the author—where the position on said problem is summarized. It began with the problems about the nature of physical bodies, which was the form of cosmological analysis par excellence in the Aristotelian tradition of scholasticism. But from there, a series of problems are included that refer more to an experimental physics than to a contemplative one: the impenetrability and porosity of bodies, their divisibility, their density and its opposite, or their *gravity* and its opposite. All these latter issues showed the path of Physics' autonomy, as the scientific advances of the time allowed to provide concrete experimental answers. And in a certain way they also showed the tensions that Physics had to go through: the clearest example is the problem of the impenetrability of bodies. This becomes a central problem of all physics (and metaphysics), because if it is possible to define whether a body is penetrable or not, we are in a position to define the minimum structure of a body, of matter. This is a problem that is dragged from the Physics of Aristotle and his discussion with the atomism of his time: what is the minimum-or basic-structure of the world? And Zavaleta here not only discusses with this Greek tradition, but with Cartesian rationalism or the empiricism of Gassendi. He addresses the problem of whether the mathematization of the Physics is an appropriate tool to solve this problem, for example, if we can consider-as in geometry-the *point* as the basic structure of physical reality, and how this influences the problem of the continuum. The crucial thing about this type of sections is that they allow us to see the winding path that a discipline should traverse to emancipate itself and more clearly demarcate its own interests from those belonging to a philosophical tradition. Abandoning the philosophical approach not only implied attending to the recent advances that were occurring in Europe, it implied transforming the problems and the entities that constituted them. One of the deepest transformation turns was to understand that the answers to these types of questions could no longer be thought of in terms of qualities of entities. This seems to have been one of the philosophical forms of thinking that most resisted before the intellectual elaboration of the new physicists.

The third section of his course was devoted to the problem of movement, and here again the influence of the most recent approaches of his time is shown. The explanation of movement no longer refers to the predicates about the nature, quality and space of bodies, but to the study of certain specific types of movement such as that of gravity, reflection or collision. The fourth section accounts for, effectively, the break with Aristotelian philosophy of physics. These are problems of a modern physics, as experimental, and refer to the issues of balance, heat, solids and fluids, taste, smell and sound, light and colors. Although the approaches are not experimental, and do not make direct reference to the most recent advances of the time, the mere arrangement of these topics as self-contained problems gives us a sample of someone close to the movement of knowledge in Physics.

We see that this course by Zavaleta from the late eighteenth century allows us to recognize that the force of scientific advances in the old world, especially in physics, quickly permeated some teaching spaces in the Rio de la Plata. Almost forty years later, in Montevideo, where there was almost no educational tradition in Philosophy, Physics showed elements that made it think with certain degrees of independence from the set of philosophical knowledge.

We have mentioned Alejo Villegas as one of the main responsible for bringing to our territory the formative experience lived in the Colegio de San Carlos in Buenos Aires. In 1839 his General Physics program is published in the newspaper El Nacional, which is recovered by Ardao in his book *Pre-university Philosophy in Uruguay.* Two years later Villegas himself proposed some changes to his General Physics course, without any substantive differences being observed from this one that was published in the press of the time. Succinctly, General Physics continued to focus on the problems of movement and balance, fundamental laws that still showed the residue that remained from Aristotelian Physics. However, as we have seen, the analysis of passive movements (inertia and gravity) were taken up from the postulates of Isaac Newton, which shows that Villegas maintains the interests of thinking about physics from modern postulates. It seems less clear the understanding of the matters included within a Particular Physics, as it required at that time to delve into the fields of experimentation. It is highly likely that these philosophers of a *physicist* tendency showed their preference towards the experimental development of physics, but that seemed very different from deeply understanding its problems and experiments in full.

In a summary summary, the continuities and small changes in the study plans of the Physics section within the philosophical training, if we compare the programs of the Zavaleta course (1795–1799) and the one taught by Villegas in Montevideo (1839). The teaching of general physics in both programs maintains a focus on the fundamental properties of matter, such as impenetrability, divisibility and density. Both Zavaleta and Villegas insist on the analysis of physical bodies from traditional concepts, with a notable continuity in the study of nature and the movements of bodies, this being a recurring theme both in the studies of classical physics and in the nascent attempts at mathematization of physics. Both programs also share a concern for the fundamental principles of movement and gravitation. In the case of Zavaleta, the movement of gravity, reflection or collision is addressed, while Villegas extends these concepts to the movement of celestial bodies, integrating the laws of Kepler and Newton, which shows a continuity in the interest for the systems of movements both terrestrial and celestial.

One of the main observable changes is the way in which particular physics is conceived in each program. While Zavaleta follows a more scholastic tradition, analyzing problems such as balance, heat, solids and fluids, from a non experimental approach, Villegas' program gradually introduces experimental physics. This transition reflects the advance in the step from a philosophical-speculative treatment, present in Zavaleta with the question about the adequacy of mathematization to solve physical problems, towards a more technical approach and based on specific laws in Villegas. This is evidenced in the inclusion of Newton's and Kepler's laws, and in the study of specific material properties such as elasticity, ductility and toughness. Villegas also introduces the study of gases, something that is not present in the Zavaleta plan, reflecting the inclusion of more diverse phenomena.

1.3 Physicists from Two Shores

The historian Guillermos Furlong has accounted for a movement, mainly centered in the University of Córdoba and the Colegio San Carlos in Buenos Aires, which-as we saw-he called *physicists*. The professors included in this category, who marked their imprint with a new philosophy, marked the trends of both institutions towards the end of the eighteenth century and the beginning of the nineteenth.

> By physicists we do not mean the supporters of physicalism or the religious system of Saint-Simon, but the philosophers who, in the course of the eighteenth century, particularly in its latter stages, granted a total, or at least excessive, preponderance to the physical sciences, to the detriment, or at least, with disregard for the other parts into which traditional Philosophy was divided.
>
> Professionally they were not physicists, but philosophers, but the fantastic development of the physical sciences drove them to study them, and their philosophical education led them to want to reconcile the results of the sciences with the philosophical principles, which they had nursed in the school classrooms. The reconciliation was very difficult, if not impossible, and the decrepit Physics of the Peripatetics contributed to discredit all Scholastic Philosophy, and to lead it to contempt, or at least to the abandonment of the philosophy they were imbued with.[10]

Among the names that stand out under this new philosophical imprint are the clerics Juan José Paso, Melchor Fernández—who was a teacher of Larrañaga-, Mariano Medrano—who also maintained in Montevideo some years of activity after the Revolution of May 1810–, Manuel Gregorio Álvarez, Valentín Gómez-who was a priest in Canelones, Uruguay-, Juan Antonio del Valle, Manuel Gervasio Gil—a disciple of the physicist and astronomer born in Ragusa, Roger Joseph Boscovich, and Alonso Frías. Many of these names, through their lives, had contact and influence on the scientific future of Uruguay, especially in physics.

The position of the *Physicists* was not only contrary to centering philosophy on metaphysical or theological disquisitions, it required a new intellectual effort, the full understanding and even the practice of certain experiments. In a way, the way in which physics develops in the context of this movement has more to do with a more rigorous attention to what was understood as Particular Physics, than to that more philosophical part as was the General Physics. But in both parts of what comprised physics, the new references marked the great break with the scholastic tradition. If General Physics was located in the long tradition that emerges with the Aristotelian postulates, in Particular Physics it sought to attend to the new experimental findings that encourage the possibility of creating new fields of knowledge within Physics.

Without a doubt, the figure of Isaac Newton is the great source from which this new way of conceiving General Physics draws. The problems about the nature of physical bodies, their solidity and fluidity, movement, inertia and balance; are addressed by the authors *physicists* taking Isaac Newton as a reference.

[10] Furlong (1947, 487).

Alejo Villegas explores the Newtonian definitions: "the passive of inertia, and the active ones of attraction and movement". The problems that the Uruguayan philosopher identifies as the main properties of matter (extension, impenetrability and divisibility), make up the central problems of the new Rational Mechanics that Newton exposes in his Mathematical Principles of Natural Philosophy (1687):[11]

> On the other hand, we, cultivating not the arts, but philosophy, and writing not about manual forces, but about natural ones, we deal above all with everything related to gravity, lightness, elasticity, resistance of fluids and forces of the like, whether of attraction or repulsion; and therefore we propose these as our mathematical principles of philosophy. For all the difficulty of philosophy seems to consist in that, from the phenomena of movement, we investigate the forces of nature and after these forces we demonstrate the rest of the phenomena.[12]

A separate chapter in the conception of Physics can be seen following the tradition of installing astronomical observatories, throughout the nineteenth century, which to some extent transforms the practice of astronomy and contributes to the disciplinary autonomy of physics.[13] Villegas cultivates, as we saw, the new understanding of the solar system from the first postulates of Johannes Kepler (1571–1630) and the corrections that Newton made of those.

The other great topic of reference in any General Physics has to do with the composition of matter: its impenetrability and its continuity. This is a topic that was still treated from perspectives strongly philosophical but in authors who were part of the more technical disputes that arise from the Newtonian advances. In this sense, the new philosophy *physicist* focuses on Newton as well as on authors who offered alternative positions, such as Gottfried Wilhelm Leibniz (1646–1717) and Ruder Josip Boscovich (1711–1787). Thus, just as it happened in Europe, the mechanistic explanation of the world that was gaining ground especially from the incorporation of the great Newtonian work, found followers to the positions of resistance that they had to do with a certain philosophical vitalism, which was part of the criticism of the English philosopher and mathematician. This criticism is recognized in the appreciation of Leibnizian philosophy, its objections to radical Cartesianism and its criticisms of the notions of movement (*conatus*) as part of a resilient metaphysics within the new philosophy. This framework of European discussions can be traced in a much more limited way in our latitudes. The case of Manuel Gervasio Gil, who-as was mentioned-studied with Boscovich, records one of the first cases of direct influence with the mathematical and physical thought of Ragusa: "The physicist of

---

[11] Although there have been no studies on the reception of Newton's work in Montevideo, the references of Furlong (1947) and Lértora (1972) provide a regional clue about this impact of the "enlightened philosophy". In more general terms, the work of Saldaña (2007) shows this impact in the Latin American context.

[12] Newton (2011, 98).

[13] By the nineteenth century, international astronomy was expanding its working universe. From being an astronomy dedicated entirely to the positioning of the stars in the cartographies that were built from different points on the planet, it began to include among its tasks the determination of the composition of these stars. See Keenan (1991) and Paolantonio (2011).

true talent and who gave enormous development to his conceptions about the experimental sciences, in greater or lesser consortium with Philosophy."[14]

Gil was born in the reign of Aragon, in 1745, and at the age of seventeen he entered the Society of Jesus. The persecution of the Jesuits forced him to be deported to Italy where he will carry out his studies in philosophy, excelling in mathematics and physics, thus obtaining a position as a professor. His passage through the Río de la Plata was quite exiguous, and his arrival is located in 1767 although his departure also seems to have been prompt. His performance as a professor places him in different cities such as Cremona, Parma and Plasencia. Father Gil writes a work dedicated to the defense of Boscovich's ideas, especially those related to the problem of the composition of matter. Boscovich tries to mediate between the two great influences of the time in this regard, the ideas of Leibniz that particles are abstract points without extension, and the corpuscular idea held by Newton, where the particles interact through forces at a distance, recognizing himself that the ideas and calculations of repulsion and attraction could serve to explain how these particles occur in matter. As some contemporary authors have underlined, Boscovich-and his disciple from the Río de la Plata-composed the first ideas about the nature symmetric of physical magnitudes;[15] which allows us to see how also the scientific and philosophical tensions existing in Europe also reached our latitudes.

As for Particular Physics, the inputs seem more difficult to trace. This difficulty has to do with the inclusion of experiments that allowed a better understanding of some natural phenomena. Although the evidence of experimental practices in the early decades of the nineteenth century seems difficult to locate, there are some indications that deserve some attention. The so-called Particular Physics gathered a series of knowledge about certain aspects that began to be treated in a, precisely, particular way.

> (...) the notable and considerable contribution of experimental physics inclined to introduce issues that strictly do not have a philosophical solution but rather a scientific one, depending very directly on laboratory experiences. Its inclusion in a Philosophy Course is debatable from a systematic point of view, and especially for placing them on the same level as the previous ones and pretending a philosophical response. However, it indicates on the one hand a commendable openness of thought by not pigeonholing in old topics, and on the other a work of information that from the point of view of teaching can be considered convenient.[16]

Within this part of the studies arise the problems related to equilibrium, that is, the question related to the points of compensation of solid bodies and liquids; to heat, no longer understood as sensation but as an objective source of bodies that is produced from the movement of particles; taste, smell and sound, trying-like with heat-to objectify the treatment that traditionally was oriented to sensations, from the identification of the physical processes that produce these sensations; light and colors, which means a great break with the peripatetic tradition, by showing that the

[14] Furlong (1947, 512).
[15] Rosochotsky (2022).
[16] Lértora (1972, 20).

composition of colors must be included in the general framework of studies on light, which was generally conceived as the emission of particles in a straight line from the luminous object; and finally, we must include the first experiments on electricity, which deserve a small section.

The Ukrainian physicist and astronomer George Gamow has called the nineteenth century, the century of electricity. It is undeniable the impact that the discoveries made in this branch of physics had in our region, which can be verified at least on two levels: on a first level, showing that Particular Physics found in the reproduction of experiments a genuine manifestation of moving away from philosophical speculations and allowing paths for experimental science. In this sense, experiments in electricity were highly valued for their relatively easy realization. But on a second level, it is observed that through these new discoveries it was achieved to perceive more latently the benefits of the practical application of electricity-and of science-in national development. It is not a minor fact that the first careers in Medicine and Mathematics included courses in Physics, and that they devoted several of their topics to the application of physical knowledge. A clear example of this is the experiment of the Leyden jar, created in 1745 by Pieter Van Musschenbroek, where electricity is stored for the first time. These experiments will be present, for example, in the first Physics chairs at the Faculty of Medicine.

1.4 The University of Montevideo in 1849

The faint tendencies towards provoking a scientific cultural development were also glimpsed in the national press of the time. Not only from the participation of these intellectuals in the most sympathetic newspapers, but in new editorial lines that showed an unusual attention to the need to focus journalistic projects based on information coming from science with an impact on the activities of Montevidean society of the time. One of these newspapers was "The Bee of Silver". This newspaper focused its concerns on the advancement of science and its application to national problems, which showed the concern of some lesser-known intellectuals, which serve as a basis for building more detailed on the history of the country's scientific thought. Such was the case of the role that Marcelino Pareja played in the publication "The Bee of Silver". In 1837 this newspaper was published, which radically differed in its editorial line from the rest of the publications of the time. When important newspapers, like "The Initiator" focused their attention on the political and cultural life of the Rio de la Plata, "The Bee of Silver" defined itself as a newspaper "dedicated to commerce, industry, education and instruction". In an editorial of the first issue, this perspective was more explicitly emphasized:

> We propose to transmit to the public the knowledge that can contribute more especially to multiply our industrial resources, and to improve our public and domestic customs. We will extract from pure sources whatever we judge capable of illuminating the use of capital,

> facilitating the operations of our industry, developing the germs of our wealth, accelerating and spreading the circulation of enlightenment, and stimulating the flight of our social progress.[17]

"The Bee of Silver" centered its contents in three large sections: an inaugural one where it sought to instruct on new applications to the country's livestock and agricultural industry; a second dedicated to the presentation of data related to agricultural and livestock trade, demography and also climate issues; and finally a section dedicated to citizen instruction, whether this domestic or national. Its responsible director was Marcelino Pareja, the first professor of Political Economy in Uruguay. In that first issue appears translated from French a text by the German scientist Alexander von Humboldt, titled "On the influence of the monastic spirit and regime on the character and social condition of the indigenous tribes of the former Spanish colonies,"[18] where the wild character of the colonial process in American lands is narrated. The attention to the scientific view of social problems will successively appear in new issues of the magazine, not only highlighting the defense of a socialist ideology defended by Pareja, but above all a methodical analysis and established based on data of the most everyday problems of social life. In this sense we draw attention to an article that begins in issue six of the publication, titled "Education and Instruction", translated from a proposal that was published in the *Journal of useful knowledge*, organ of the *Société nationale pour l'émancipation intellectuelle* of France, which was created by the journalist, writer and politician Émile de Girardin. In it not only appears a proposal on how to organize the education and public instruction of children and young people, there clearly appears a reference to the place of physics in such training, and even more, the need to have scientific societies that drive a new change in citizen education. Two are the elements that stand out in this proposal, which Pareja decides to translate for the Rio de la Plata context: the idea of a generalization of teaching "Experience shows that when elementary instruction is the privilege of some instead of being the obligation of all, it only makes victims;"[19] and the construction of it based on an intellectual literacy that incorporates the scientific knowledge to the manufacturing and industrial use. In this last sense, the idea is recognized that at a second level of initial training, children should be instructed in" Notions of Physics and Natural History", being this proposal one of the pioneers that reach our lands. And in the same line, a footnote is highlighted in which the newspaper itself, perhaps Marcelino Pareja himself, makes some annotations regarding the role that new artistic, scientific, literary societies, could fulfill to achieve the purposes set out in the article:

> Literary, scientific, artistic, moral, charitable societies, there are the powerful levers that have removed the enormous obstacles that the concerns of previous centuries had accumulated, to prevent the progress of reason and the empire of truth. Before the wonders of natural sciences, which wise bodies cultivated since the beginning of their institution, fell to the

[17] Pareja (1837, 1).

[18] von Humboldt (1959 [1810], 9–10).

[19] Girardin (1837, 66).

ground the simulacra of pedantry, fled from the world in which the genius of error dominated, and the humble of the earth finally knew the beneficial truths to which they owe the improvements, always increasing, of their social condition, and all the means of their domestic well-being. Such has has been the influence of learned societies on the destinies of the world, and such must still be, according to our way of seeing things, that we do not judge risky to assert that, without them, it is entirely unverifiable, in any country, and whatever the degrees of its social advances, all intellectual progress in the masses, every important discovery to the arts, every fixed direction to customs, and every favorable avoidance to the public spirit.[20]

At the base of this context is the creation, in 1849, of the University of Montevideo, in a process that has already been detailed in authors like Blanca Paris (2010). Manuel Herrera y Obes, member of the government of Joaquín Suárez, was the main actor in promoting this installation along with the person in charge of the Gymnasiums that had been operating for a few years, Luis José de la Peña (who had also been formed under the scientific influence of the San Carlos College of Buenos Aires). Both led the purpose of promoting a university that on one hand, mitigate the ecclesiastical formation that was imparted in Montevideo since the colonial period; and on the other, make viable through education the new independentist project of the nation.

That is the idea that fundamentally presides over the installation of the University of Montevideo [...]: the idea of achieving, once political independence has been consummated, the mental independence of America; to free the country through education, from the political-social habits that. The afflicted as a legacy of colonialism.[21]

In the installation records of the University of Montevideo, the first problems faced by the new institution begin to be recognized. It should be remembered that in the installation of the University it is entrusted to "embrace" all public education, which was divided into *Primary education*, *Secondary education* and *Scientific and professional education*. This last training was divided into four faculties, which followed the Germanic model of organization: Faculty of Medicine, of Jurisprudence, of Theology and of Natural Sciences. In the Faculty of Natural Sciences it is proposed to attend: transcendental mathematics, drawing and its applications, and the principles of agriculture, botany, chemistry, navigation and architecture.

The organization of physics was contemplated from the beginnings of the University, in a course that was called physical-mathematics, inspired by that part of philosophy that in 1837 was carried out by José Benito Lamas. This course was divided into a two-year training, where in the first one instruction was given in the problems of arithmetic, algebra, second degree equations and elementary geometry. The second year would teach rectilinear and spherical trigonometry, practical geometry and general physics. In a way, at least from a formal perspective, Physics becomes a separate study from Philosophy. We say that it occurs on the formal plane because there were several problems to find teachers for this chair-which in fact was taught by the professor of Philosophy-, which was for a long time without a

[20] Pareja (1837, 94).

[21] Paris (2009, 20).

reference. Alejo Villegas, member of the first Doctors' Room, was assigned to the course of Canon Law. In 1851 it was reported that the chair of physical mathematics had 125 students (out of a total of 457, which made it the second most attended chair. Even so, it was reported that

> The chair of Physical-mathematics has not been provided with a permanent professor: it has been performed for 2 years by the professor of philosophy, who is therefore under extraordinary work[22]

For this and other reasons, the Doctors' Room recognizes that

> The progress of studies indispensably demands that the lessons given by the professors in their respective classrooms be perfected by reading scientific works of all kinds. This need gave rise to the thought of creating a particular library for the University: and it has begun to be realized through donations being the first in them the members of the University Council (…)[23]

On March 16, 1852, the University Council receives the proposal to appoint Dr. Alfredo (Du) Pasquier in the chair of Physical Mathematics. Pasquier seems to have come to our lands through his Masonic contacts, having trained with a degree in Mathematics and a Bachelor's degree in Letters at the University of Montpellier.[24]

> Alberto Larroque (lawyer and former teacher at the University of Montevideo) invites him to the Argentine city of Concepción del Uruguay to teach at the Colegio del Uruguay, the first secular school in Latin America, where he taught Physics (…)[25]

Pasquier's work will be fundamental in the following decades for the advancement of mathematical thought that would consolidate in 1888. with the creation of the Faculty of Mathematics. Especially his contributions in studies on algebra will be definitive for the consolidation of a metric system in Uruguay.

Later, in the university records we know that Dr. Adolfo Pedralbes was also a professor of Physical Mathematics until 1856, he also integrated the Metric System Commission, which aimed to unify the system of weights and measures of Uruguay. Once he presented his resignation in that same year, Dr. Antonio Dupart took over, he also privately taught contents of mathematics, physics, architecture and civil engineering, after his resignation in 1858, Martin Pays took over.

In this same logic we must include the presence of the philosopher French Amadée Jacques in his passage through our lands and his attempt to install a physics cabinet at the University. His arrival to our lands, in 1852, should not be thought of as an isolated milestone. Let's imagine that after almost a hundred years of established European colonies in this region, the emergence of the Creole class (second or third generation of those conquerors who settled), began to erect the thought and political action for the independence of the territories. This was not only a scenario of possibilities for the small group of Creole intellectuals, but it also meant an open

[22] Minutes of the University of the Republic (1849, 97–98).

[23] Minutes of the University of the Republic (1849, 99).

[24] Minutes of the University of the Republic (1852, 118).

[25] Rodríguez and Martínez (2020, 23–32).

field for the intellectuality-especially French-disenchanted with the design that Europe had acquired after the French revolution. For many French people, the possibility of emigrating to America was also the possibility of founding a new social project free of pernicious traditions. This seems to be the fate that befell Amadée Jacques, who arrives on our shores with a suitcase, some chemistry and physics equipment, and a letter of recommendation from Alexandre Von Humboldt:

> I recommend to all those who in the beautiful regions of the South America have preserved the memory of my name and the good reception of my works, the bearer of these lines, Mr. A. Jacques, a writer as distinguished by his talent as by the great variety of his knowledge. Linked for many years by the most affectionate ties to the respectable family of this scholar, I took the most lively interest in his fate and in the execution of his project to found in the New World, in the center of an active population, an establishment of public instruction. Mr. Jacques is in a position to achieve such a noble purpose, not only because of his culture and the elevation of his character, but also because of the experience he has acquired in the honorable positions he held in France, as a professor at the College Louis the Great and as a master of conferences at the Normal School. The services that may be rendered to him will be for me, the oldest of the travelers of America, a motive of the most lively recognition. Professor Jacques has long been entitled to the high esteem that I have consecrated to him.
>
> Alexandre Von Humboldt
> Berlin, May 1, 1852.[26]

His training and professional performance was in the field of philosophy, having moved in France from eclecticism towards a clearly positivist perspective. But he also had some experience in the experimental field of science having done practices in chemistry laboratories with Auguste Laurent, on organic matters. In addition, he had worked in the family business of the glassware. Therefore his instrumental material was of great impact upon his arrival at the university: his physics cabinet consisted of equipment that allowed experiments to verify certain natural laws and for meteorological measurements, a microscope and a device to produce electric light, relief figures for the study of geometry and scale models of machinery: hydraulic wheels, mills, tripods, cranes, plows, steam engines, rolling mills, turbines and presses. In a letter from 1853 he writes to a friend who lived in Paraguay, telling him that his course had several students, whose interest was more oriented towards the instruments than the reason behind them. In that same letter he confesses his plan not to practice philosophy and dedicate himself to teaching sciences, which would not be difficult for him:

> It would be useless and almost ridiculous to bring metaphysical speculations to a so new country, for which the most convenient thing is a practical teaching, above all, of the sciences applied to agriculture, to commerce and industry (...) Easy because the teaching of philosophy, although different from that of the positive sciences, was in France linked to these, first of all by their very nature, then by the requirement of certain common graduations to the two faculties, which I had acquired; pleasant because I have always had a decided predilection for the natural sciences and their applications, which I have cultivated constantly"[27]

[26] Vermeren (1998, 90–91).

[27] Vermeren (1998, 92).

Despite having called about a hundred students, the teaching of the course at the Major University will not prosper because he will not be paid for his tasks. This seemed to be a very common situation, manifested in several discussions of the University Council especially from the end of the Great War. Amadée Jacques must leave Montevideo, and it is again Dr. Alberto Larroque-who had called Alfredo Pasquier-, who takes him to the school in Paraná where he directed the institution. There ends the dream of forming a professional school or polytechnic institute in Montevideo:

> (...) a place of teaching physics, chemistry, mechanics, mathematics, geology, not only theoretical, but also practical, to know the thousand useful inventions to satisfy the needs, increase the forces, fill the ignorances; to use the natural forces and ward off their danger, distinguish the good things from the bad, to separate deceptive mixtures, to test natural substances and artificial products, to know how to locate a mineral deposit and discover its nature and wealth, to keep a record of expenses and trade books, to measure a field or formulate a plan, etc... An establishment that would have wanted to create completely outdoors, in the countryside, in some clear and healthy mansion, with a body of young and active teachers coming from France, not pedants willing to participate as much in the students' games as in their work, and where nature itself, where real machines would have replaced the reduced models, where iron, stone and wood would be worked under the direction of skilled workers, where the latest European methods in agriculture would be experimented with in practice. An institution, in short, that would convene. Students from all corners of the Rio de la Plata, and even from Chile, from. Peru and from Brazil. The State would support this project that would illuminate the beacon of material civilization in South America.[28]

Bibliography

Ardao, A. 1945. *Pre-University Philosophy in Uruguay*. Montevideo: Claudio García & CIA Publishers.

De Asúa, M. 2010. *The Science of May. The Scientific Culture in the Río de la Plata, 1800–1820.* Buenos Aires: Fondo de Cultura Económica.

Duffau, N. 2016. Population and society. In *Uruguay. Volume I—1808/1880: Revolution, Independence and State Construction*, ed. G. Caetano, A. Frega, and M. Etchechury, 175–221. Montevideo: Editorial Planeta, Fundación Mapfre.

Furlong, G. 1947. In *Birth and Development of Philosophy in the Río de la Plata 1536–1810*, ed. G. Kraft. Buenos Aires.

Girardin, É. 1837. Education and Instruction. Translation by Pareja Marcelino. *The Bee of the Plata* 1:6.

Keenan, P. C. 1991. The earliest national observatories in Latin America. *Journal for the History of Astronomy* 22 (1): 21–30.

Lamas, J. B. 1836. *General Description of Public Acts of Conclusions and Exams: of the Philosophical Course.*

Lértora, C. 1972. The course of "General Physics by Diego Estanislao de Zavaleta". *Cuyo* 8:First era.

[28] Vermeren (1998, 93–94).

Minutes of the University of the Republic. 1849. *Functions and Organization of University Studies*. General Archive of the University of the Republic.

Minutes of the University of the Republic. 1852. *Report Two Years After the Foundation*. General Archive of the University of the Republic.

Moreno, M. 2004. *Life and Memories of Dr. Mariano Moreno*. Buenos Aires: Historical Bookstore.

Newton, I. 2011. *Mathematical Principles of Natural Philosophy*. Madrid: Alianza Editorial.

Pareja, M. 1837. Education and Instruction. *The Bee of Silver* 1:8.

Paris, B. 2009. *The University of the Republic. In the Formation of Our Liberal Conscience*. Montevideo: University Editions.

Paolantonio, S. 2011. Los inicios de la Astrofísica en Argentina I. *A partir de la ponencia realizada en el Encuentro Internacional Pro-Am LIADA/XIII Convención de Astrónomos/II Simposio de Astrofísica/LIADA*, Santa Fe.

Rodríguez, W. A. M., and P. N. Martínez. 2020. Teaching of acoustics in Uruguay. Origins, facts and protagonists until 1930. *ECOS* 1 (2): 23–21.

Rosochotsky, A. 2022. RJ Boscovich on physical symmetries. *Studies in History and Philosophy of Science* 93: 149–162.

Saldaña, J. J. 2007. *Science in Latin America: A history*. University of Texas Press.

Vermeren, P. 1998. *Amadeo. Jacques: The Democratic Dream of Philosophy*. Buenos Aires: Ediciones Colihue SRL.

Von Humboldt, W. 1959 [1810]. About the internal and external organization of higher scientific institutions in Berlín. In VV.AA *The idea of University in German*. Editorial Sudamericana, Buenos Aires.

Chapter 2
Towards Scientific Faculties

From a comparative perspective, the creation of the Faculties of Medicine (1875) and Mathematics (1888), which will contemplate in their teaching some aspects of physics, may seem a path towards the autonomy of the discipline. In a way, the canonical bibliography on the history of Udelar seems to want to show that the reforms promoted by Alfredo Vázquez Acevedo[1] led to transformations that, among other things, sought to adapt "to the scientific progress of the time."[2] On the other hand, Blanca Paris characterizes the same period using expressions from Eduardo Acevedo:

> The regularity and breadth of the courses will no longer be disturbed by the deficiency of the classrooms and the scarcity of teaching materials; and good teaching methods will no longer be contained by the lack of apparatus and scientific collections. I attribute to all this the greatest importance. Learning in almost all sciences cannot be well achieved with the simple aid of books. Students do not manage to realize the laws that govern physical and natural phenomena; and the great efforts they necessarily make to meet the demands of their teachers, end up arousing boredom or deep reverence for study.[3]

In this way, it is often argued that Uruguay and its university would have entered a stage reigned by "scientism", as material resources were provided to strengthen physics or chemistry courses, with appropriate equipment to accompany teaching. While this can be analyzed in a general and very simplistic way as a transformation of the university turned towards a better attention to scientific knowledge, such transformation does not enclose what essentially defines scientific universities. And

---

[1] Alfredo Vázquez Acevedo (1844–1923) was a jurist and writer born in Argentina and based in Uruguay. He was rector of the University in the periods from 1880 to 1899, with two interruptions, in which he carried out a reform on the positivist view with which the ideological thought of the University of the last decades of the nineteenth century is usually characterized.

[2] Ardao (2008, 77).

[3] Paris (2009, 67).

J. A. Queijo Olano, E. Mozo Meneses, *Giants Upon Our Shoulders: A Philosophical History of Physics in Uruguay*, Studies in History and Philosophy of Science 65, https://doi.org/10.1007/978-3-031-97729-9_2

the distinctive element is that, despite the resources poured, the ultimate concern was in the teaching of the student, and not in research. It is paradoxical to see how, in the same historical period, in Germany the concern was centered on the need for the state to allocate its resources towards laboratories and institutes so that, precisely, they do not limit themselves to the function of teaching.[4] But while spaces dedicated to research in these new faculties created are not recognized, a closer look allows us to recognize academic trajectories where there is a different concern regarding the teaching function reduced to its pedagogical function.

The chair of physical-mathematics remains fundamentally unaltered during the period that spans from its installation, a few years after the establishment of the Universidad Mayor de Montevideo, until the formation of the two new "scientific" faculties. It will only be from there that we can recognize a significant change in the conception of physics that until then was still markedly defined by philosophical approaches. The projects of the Faculties of Medicine and the Faculty of Mathematics will show new ways of understanding physics, in particular, and a new conception of knowledge in teaching and in society.

Without a doubt these new faculties condense national needs, intimately associated with projects for the development of the country: on the one hand, a development towards a modern health system, and on the other, a development of public works (mainly roads) that will improve the territorial communications of a small country but extremely poorly connected. In this sense, the institutional construction of both Faculties imprints a technical substrate to the country until then only developed in jurisprudence, whose purpose does not stray from the purpose that had always been given to the University: to serve in the formation of the ruling elites. Now, these elites will also be technical, associated with the training of doctors and engineers.

As was remarked, while the changes that occurred in the rectorates of Vázquez Acevedo must be recognized, they do not seem to be decisive in the properly scientific development of the country, at least not in the branches of knowledge included in the basic sciences. And this is not a minor thing to remark and to slightly shade the narrative that historically has proposed Ardao. At the time of these transformations that are driven with Vázquez Acevedo, it is worth remembering at least three relevant elements: first, the situation of scientific development in Europe, in knowledge centers with some connection to our country; second, regional experiences that did mark a significant contribution to the development of science from these latitudes; and third, the work of some individuals within the University who attended to the problems of physics seeking to come into contact with the most updated advances of the discipline, in approaches to scientific advances.

While the role that Dr. Eduardo Vázquez Acevedo played to favor the conditions of a university transformation that enables a better entry of the sciences into the disciplinary set of knowledge offered by that house of studies is undeniable; it is not minor to show that there also existed, prior to his rectorate of 1880, some figures

[4] Cahan (1985).

who maintained a close link with the scientific novelties of the time, and who sought to translate them and think about them for the situation in Uruguay.

2.1 The School of Medicine (1875)

Dr. Plácido Ellauri (1815–1893), rector of Udelar in two periods (1871–1873 and 1875–1876), was a central figure in the foundational stage of the Faculty of Medicine. First, during his first rectorate he proposed a reform that included a budget for a chair of physics and another of natural history. Then, in his second term, Ellauri strived to realize his project for the Faculty of Medicine. At the end of 1875, he convened a group of national doctors, responding to the initiative of the Uruguayan doctor Manuel María Espinosa and in the minutes of these sessions prior to the establishment of the Faculty of Medicine, the project of regulations developed by the commission that will be approved by the Government, constituting the first code for the House of Medical Studies. Thus, the first two chairs of Anatomy and Physiology were set in motion:

> Faculty of Medicine, Decree of its foundation
>
> Ministry of Government.
>
> Montevideo, December 15, 1875
>
> The President of the Republic has ordered that the chairs of Descriptive Anatomy and Physiology be installed in the Major University, of the Republic and in the next year, for which the University Council, in agreement with the Health Board, will appoint the professor or professors who must perform them; this appointment must be submitted for the approval of the Superior Government.
>
> It is also communicated to you that for the establishment of the referred Chairs, it has been ordered that the budget of the University be increased by four hundred pesos per month from the next month of January. (...)
>
> The Government will dedicate preferential attention to fulfilling as it should the needs of this establishment. God save the Council.
>
> Tristán Narvaja[5]

With the founding decree issued on December 15, 1875, the Faculty of Medicine was established, beginning its operation in the next year. The Government allocated additional resources to the university budget for this purpose. Once the chairs were established, the creation of the remaining ones necessary for the training of the doctor proceeded. For the interest of this work, the analysis of the chair of Medical Physics and of its main representative: Jacinto de León seems important. The creation of this chair where physics found some developments is not merely anecdotal. Contrary to a widespread opinion, the institutional links between the disciplines of medicine and physics were not born with the radiotherapeutic techniques that use Röntgen's discoveries, on the contrary, a whole previous tradition in medical physics had developed as a field in Europe.

[5] Herrera and Gorlero (1959, 121).

> The first use of the term *Medical Physics* (or to put it more accurately, *Physique médicale*) was in Paris in 1778. The term was introduced by the general secretary of the *Société royale de médecine*, Félix Vicq d'Azir (1748–1794). Physics was explicitly included in the work of this society alongside the other basic sciences such as botany, natural history and chemistry, and reinforced in the title of its journal, *Les Mémoires de médecine & de physique médicale*.[6]

Medical physics involved, in its first trials, the extension to the field of human anatomy and physiology of the general principles that were beginning to be discovered in the world of natural physics. In this sense, it is not surprising that the first globally characteristic investigations of this disciplinary field include the therapeutic studies of electricity. The famous experiment of the Leyden jar, carried out by the Dutch physicist and doctor Pieter van Musschenbroek in 1746, was a source of inspiration for animal experimentation with electricity. Musschenbroek's experiment was one of the few that were attempted to replicate in the chairs of medical physics at the University, and as we will see later, De León was a follower of these physical applications in medicine.

French medicine will be a precursor of this relationship between physics and medicine, in politically turbulent and revolutionary times. Antoine Lavoisier himself, imprisoned awaiting his execution, recognizes that the destiny of the physicist should be guided by providing solutions for the improvement and increase of human life.[7] In any case, the idea that medicine should be practiced on the solid basis of knowledge of physics was widespread among the practitioners of French medicine, which was of great influence for the development of this art in Uruguay.

2.2 Medical Physics: José Jacinto De León

Jacinto de León was born on April 7, 1858, in Tala, department of Canelones, Uruguay. Once he completed his teacher training, he taught botany at the University High School while simultaneously studying at the new Faculty of Medicine. He traveled to Naples in 1881 once he finished his medical training with the purpose of expanding his medical knowledge. A few years later, in 1883, he returned to Montevideo and graduated with the title of surgeon doctor, being recognized as the first Uruguayan neurologist. Between 1886 and 1887 he participated in the intense political agitation of the country, contributing with funds for the armed protest that took place by the opposition of the then elected president Francisco A. Vidal, resulting in exile for a few months in Argentina where he provided medical services to the Public Assistance of Buenos Aires, during a cholera epidemic. Upon his return from his exile in July 1887, he was appointed professor of medical physics, holding the chair for 29 years.

---

[6] Duck (2014, 399).

[7] Duck (2014, 300).

As previously mentioned, the purpose of medical physics is to understand the main diagnostic and therapeutic applications of different physical phenomena in the body. Jacinto de León dedicated himself especially to the study of the possible physiological applications of electricity, gaining experience in the use of various methods of electrotherapy, being a pioneer in our country in the study and practice of physico-diagnosis, which would later be known as clinical neurophysiology and physiotherapy and later will become physiatry. His contribution to the development and teaching of a physics linked to medical knowledge and with an applied purpose, is manifested from the contents on medical physics that he taught in his classes, as well as in the programs and annual plans of the memories of the Faculty of Medicine and in different publications he made on medical electricity in university journals.

From the end of the 19th century onwards, physics was a mandatory element in the contents of medical education in several European countries such as the United Kingdom, and Jacinto de León was one of the references in Uruguay in recognizing the growing importance of physics in medical practice. In the various publications mentioned earlier, his interest is shown not only in the application of physics to medicine but also in the theoretical and historical aspects of electricity, which he dealt with in the lessons of his courses.

Among the contents taught in his classes, De León taught with demonstrations the different experiments that helped to understand this phenomenon in history, introducing the outstanding "electricians" of History such as Galvani, Volta, Fabroni and the physiologist Claude Bernard. Thus, we find in his publications content on the discussions that these physicists had about the behavior of the electrical phenomenon and its nature. One of the experiments in the history of electricity that was reproduced in his classes was the famous frog experiment, a version of the famous experiment that was the beginning of discussions between Galvani, Volta and Fabroni, which led to recognizing the existence of animal electricity, propose the idea of constant electricity, and continue with investigations to know the nature of that "fluid" electricity.

> A very ingenious experience of the great physiologist Claudio Bernard gives a synthetic idea of the action of constant electricity in living organisms: suppose an electric circuit, like the one you have drawn on the blackboard, with ten or twelve batteries, gathered in tension, and in the slow interrupter circuit, a voltmeter with acidulated water, and a frog prepared a la Galvani. When the current passes, the water in the voltmeter decomposes into its elements, hydrogen and oxygen, and the frog remains still; and when the current is interrupted, opening the circuit, and when its current is initiated or its circuit is closed, the frog's leg shrinks, a jolt occurs, and the water in the voltmeter does not decompose: in the first case, there is electrolysis, and in the second a muscular contraction at the opening and another at the closing of the current.[8]

During the final decades of the nineteenth century, the status of physics in medical education becomes more relevant, mainly in its search for applications derived from extensive research on the mechanical, thermal, electrical, optical and acoustic processes of the body. Jacinto de León in his lessons on Medical Electricity, delves into

[8] De León (1905, 69).

the theoretical aspects of electricity, recognizing that at that time there was no consensus about which theory best explains the nature of electricity, but that did not prevent the development of different diagnostic and treatment applications with said agent in the field of medicine.

> We do not know the intimate nature of electricity, but we know that it is a manifestation of universal energy, it is a force, it is heat, light, mechanical work, the movement that manifests itself in that particular form that we call electricity. The study of electricity, from the medical point of view, is of great importance and of much utility: a whole group of diseases, those of the nervous and muscular systems are susceptible to being treated by means of this physical agent.[9]

Another of the contents of electricity present in the courses and interests of De León is atmospheric electricity, known as Franklinian electricity, in honor of Benjamin Franklin who demonstrated the identity of this electrical variety from the experiment that attracted the lightning from the clouds through his historic kite. The importance that De León gives to the knowledge of Franklinian electricity has a practical correspondence, because this type of electricity allowed the development of static electromotors, known as electrostatic machines or simply electric machines, which produce electricity by friction and influence. De León brought to medical practice the reaches of these natural phenomena in a medical treatment called franklinization or electric bath.[10] Another of the knowledge that he considered indispensable to understand the operation of the motors and electric generators that are used in the field of medicine was the general principle of electromagnetic induction of Michel Faraday, where the hypothesis is posed that a current can induce another current only while it is varying. The generator and the motor together made possible the electrification of many tasks resulting in a major transformation in medical practices and in all of society:

> The application of inductive or faradic electricity is called faradization. (...) Faradization can be limited to a to a member or a specific region, or generalize to the entire skin surface, in which case it is called: dry faradic bath. A simple faradic bath or faradic hydroelectric bath is the application of a bath with faradically electrified water: in this case, the electrodes are made up of large carbon plates, placing one on the surface of the water at one end of the tub, and the other at the other end and at the bottom; in addition, the tub should not conduct electricity, it should be made of wood, stone, marble or porcelain. The application of the electric hand consists of giving the patient one pole, the hollow metal cylinder in communication with a rheophore, with the doctor holding the other pole with his left hand and gently applying his fingers to the region that needs to be faradized. The duration of faradic applications varies according to the purpose: long ones of five, ten and up to fifteen minutes; when they need to act on sensitivity, they should be short, one or two minutes on each muscle when it is intended to awaken or activate movement.[11]

On this occasion, De León implements what is known as faradic or electrotherapy devices as healing agents, once again demonstrating his interest in applied physics

---

[9] De León (1896, 2).

[10] De León (1899, 276–278).

[11] De León (1899, 7).

knowledge. Developing the most varied electrical applications in the body in order to diagnose or medically treat. Therefore, he is also a precursor in the technological domain and the teaching of radiology in the Faculty of Medicine. The content he taught in 1899 shows that he was up to date with the knowledge production of the time, bringing the Uruguayan academic community closer to Röntgen's discovery when he circulated radiographs of his wife's hand among scientific colleagues in 1896, paying attention to the medical potential of this phenomenon, resulting in X-rays being quickly incorporated into therapeutic use.

> Röntgen's X-rays, which are produced when cathodes are stopped by the anticathode wall, in addition to the brilliant properties noted by their discoverer, are not deflected by a magnet, nor by an intense magnetic field, they propagate in a straight line, they do not allow themselves to be reflected, nor refracted, they do not produce thermal effects, and they pass through many bodies opaque to light, such as paper, skin, wood, aluminum, ebonite, celluloid, coal and liquids; metals are opaque to them, and bones are crossed with difficulty. In general, the resistance of bodies to be crossed by these new rays is directly proportional to their density. Röntgen rays easily impress photographic plates: this is the basis of radiography.[12]

However, in our country's Faculty of Medicine, the Institute of Radiology was only founded in 1913 by Professor Carlos Butler. Until then, this type of knowledge was only conceived as a mere novelty that only De León knew how to disseminate. The interesting thing about De León's case is that, contrary to what prevailed as "scientism", that is, the application of techniques and experimentation that reinforce university teaching, this doctor's trajectory shows us someone imbued (with all the difficulties inherent in the country's situation) in concrete research on problems of physics. It is true that his ultimate interest is medical application, as well as the university training of future doctors. But it was necessary to fulfill this ultimate goal, that the doctor be above all a physicist, who knows the theoretical bases and the experimental demonstrations of phenomena such as electricity, in order to then apply it in the professional practice of medicine. In a way, De León's trajectory exceeds the canonical narrative about the university in this period, as we see in his courses and publications the genuine research work, a word absent until now in the academic practice of the University at the end of the 19th century.

2.3 Faculty of Mathematics (1888)

The creation of the Faculty of Mathematics and Related Branches, develops as a project that better addresses the concern of the related branches than the cultivation of mathematics itself. The related branches will be engineering (of roads and bridges, mainly), surveying and architecture, for which mathematics will be required in an entirely practical sense. And in this project, physics will appear with the same

[12] De León (1899, 245–247).

purpose. As highlighted by María Laura Martínez, these concerns are framed in the processes of industrialization and national development that characterized the time.

> The Faculty of Mathematics and Related Branches, created by law in 1885, was to be a center for the training of professionals with a critical sense of national reality. It was intended that they be prepared not only technically but also committed to the country, to carry out a project of nationalization. The environment encouraged young people to take other paths that were not those of traditional careers (medicine and law), but to train in the new paths that the University opened and that the transformation of the country required; the new technical careers that responded to this first stage of modernization. The plans of the Faculty of Mathematics were organized in order to satisfy the scientific and technological needs that the incipient industry demanded, and the first young people who took them graduated from them trained not only technically, but also in the human and social aspects.[13]

Just three years later, in 1888, the courses corresponding to the engineering degree began, where about twenty students started their courses, among them Eduardo García de Zúñiga, whom we will discuss in later sections of this work. Another student from that generation, the engineer José Serrato, remembered the precariousness with which the new university project of the Faculty of Mathematics began, when the 50th anniversary of its installation was celebrated. In that speech, a phrase is collected that really evidences how unreceptive Uruguay was for training in technical and scientific areas.

> I remember with deep reverence, the years spent in the classrooms with those noble companions [...] the conspiracy of those who did not believe in the necessity or the possibility of such studies being carried out in the country; and finally, the crushing indifference, of the majority. There was no clear and defined concept in the country of what the engineering profession was. The idea prevailed, however, that it was a technician, not yet necessary, in charge of solving constructive and scientific problems.[14]

Physics will also be part of the courses that will make up the degrees of the new Faculty. Despite not having generated professional practitioners in all those decades since the creation of the Universidad Mayor, physics had a social value that allowed it to survive in a few courses within the Faculties oriented to practical purposes. This had happened in the case of the Faculty of Medicine, and so it happened in the new Faculty of Mathematics. It is interesting to see that, even from the narrative of scientific experiences coming from other places, which even-as Serrato pointed out-provoked the generalized disbelief that it was possible to develop them in our country, either because of a status or a scientistic ideal, physics always managed to be considered relevant at any stage of the construction of the University. This social legitimacy of the discipline, which was much talked about and practiced very little, is clearly illustrated in an anonymous article from 1873 in the magazine La Acacia, a weekly newspaper of the Montevidean masonry, titled "Relationship that natural sciences have with each other, and with the welfare of society."

> Ask the physicist how he has obtained such brilliant discoveries, and he will answer that his weapons are to experiment, measure and calculate; that he does not know a fact well until

[13] Rodríguez and Martínez (2020, 29).

[14] Serrato (1938, 12).

> he has measured it; that the progress of physics is in proportion to the calculation that is used; he will tell you that when he cannot measure directly he used ingenious devices, and he will show you how he has managed to measure the astonishing speed of light without leaving a small room; the millions of vibrations that the particles of light make in a second; how many back and forth turns a string of a musical instrument makes; how the caliber of a tube the thickness of a hair is measured; how the diameter of an almost invisible wire is measured, and how he has built machines to divide a centimeter into a thousand equal parts.
>
> And what is that for? some ask who look at these facts as despicable. And what is a newborn child for? Franklin answered when such questions were asked of him. And answering directly, he will tell you that it serves to broaden the limits of intelligence, to sharpen the wit and teach it to overcome difficulties that seem insurmountable (...)[15]

The social value of physics did not seem to be alien to the Uruguayan elites, at least it did not seem to be completely extraordinary to understand the cultivation of its discipline to "broaden the limits of intelligence". On an abstract level, the importance of developing physics and other sciences was clear. But it was always the pragmatic plane of science that prevailed in the inclusion of basic disciplines in university degrees. Later, the same text points out:

> The character of natural sciences is to be essentially industrial. Industry, the basis of modern civilization, a source of immense wealth, is nothing more than the conquest of matter by science. The progress of these determines the growing development of the former and all have more or less immediate applications and assist each other.[16]

This will be the characteristic that will mark the courses and approaches of the physics that will be taught in the Faculty of Mathematics and Related Branches, in charge of a true practitioner of the discipline as was Claudio Williman.

2.4 Applied and Industrial Physics: José Claudio Williman

Williman was born in Montevideo, the son of Galician immigrants who had settled in the territory. Although his political actions, linked to the Colorado Party, have highlighted him as the president who led the country between the two periods of José Batlle y Ordoñez, it is his teaching work in the teaching of physics that we are interested in highlighting. It begins when Williman, at 19 years old, teaches Physics classes at the University Society. This society brought together, under a positivist spirit, young people from the Montevidean elite who gathered around conferences, the publication of the Journal of the University Society, and courses that were provided by the Society itself.

In the presentation of this course, Williman gives a historical overview of the origins and developments of the various branches of physics that would make up the different units of the course. Thus, we have problems such as pneumatics, acoustics, heat or electricity, treated from a historical point of view, as a collection of immense

[15] La Acacia (1873, 10–11).

[16] La Acacia (1873, 30).

small discoveries. But unlike De León, Williman explicitly states that theories will not be the subject of analysis in the course, as in the case of electricity:

> I will not consider theories about electricity because it is part that does not correspond to us, I will only say that in it the experiments are not so decisive. Given that the theory of imponderable fluids is impotent to explain physical phenomena, what is their cause according to modern theories? For modern science all physical phenomena are due to movement, and as movements can be transformed into each other, phenomena should also be able to be transformed; it simply explains the facts that we have previously cited to oppose the previous theory; as according to this system everything is movement, the laws of dynamics are made applicable and mathematics come into action.[17]

It is significant to highlight that Williman's approach to physics is a good example to better understand how positivism was practiced in the Uruguayan context. Far from being a drive for science, it was rather the enlightened but superficial study of the history of discoveries, showing and highlighting time and again the benefit of scientific knowledge for its technological applications. If we make these considerations it is because the story of historians like Nahum and Barrán[18] or Caetano,[19] can lead to confusion about what that valuation for science in the country really meant: used as support for an encyclopedic teaching, detached from research and weighed for its industrial applications.

In a new installment of the same year, Williman writes about applied physics in the magazine of the University Society, returning to the topic of electricity. It is interesting to compare this perspective with the practiced by Jacinto De León, much more focused on the natural properties of electricity than Williman seemed to be:

> What best demonstrates the progress made by this branch of the physics are the series of exhibitions held with the aim of exposing the numerous electrical devices that are invented every day. Among these exhibitions are: Berlin in 1879, Paris 1881, Munich 1882, Vienna 1883 and London and Philadelphia; the latter is taking place at this moment.
>
> In all these exhibitions, the good tendency of the exhibitors to dispense with devices whose value is merely theoretical, and to deal with those that have practical applications, fully introducing electricity into the industrial domain.
>
> One of the applications of electricity that can produce a social revolution from an economic point of view, is the use of this agent as a producer of motive power and as a means of transporting mechanical force.[20]

As was commented, these experiences in the University Society will serve as merits to occupy the chair of applied physics of the recently created Faculty of Mathematics, in a program that adjusted the knowledge of physics to an application oriented to industrial development. We saw that the entire organization of the courses made by Williman had as its axis the uses and applications of the emerging areas that made up the studies of physics. Pneumatics, acoustics, heat, electricity, are the series of issues that Williman includes in his programs, but always directing knowledge to its

[17] Williman (1884a, b, 111).

[18] Barrán and Nahum (1979).

[19] Caetano (2007).

[20] Williman (1884a, b, 290).

productive application, taking mainly French bibliographic references and using concrete examples. It should not be surprising that, a few years after the Faculty was created-in 1895-, the name of the subject *applied physics* became *industrial physics*, in a Faculty of Mathematics mainly oriented to the training of surveying engineers and architects.

Despite this, Williman was a professor of physics, that is, someone concerned about the discipline as a whole, although in the training of his students he always highlighted its applications. The thematic fragmentation on the one hand, as well as its applied orientation, does not inhibit some reflections on the underlying unity of Physics. It is worth as a mere example, to return to a quote that Williman introduces from the Irish physicist John Tyndall:

> Nature is not a set of independent parts, but an organic whole. Open a piano and sing; there is a certain string that will respond to you. Change the tone of your voice; the first string will stop vibrating, but a second one will respond; modify your tone once more: the first two strings will be silent and a third one will respond.
>
> We have to, by modifying the tone of your voice simply change the form of movement communicated to the air by your vocal cords; each of the strings responds to each of the forms; in the same way the intelligent man is warned by nature, because the optic nerve, the acoustic and the others of the human body are other so many strings differently tuned and responding distinctly to the different forms of universal power.[21]

2.5 The Idea of Science in the Educational Reform of the Early Twentieth Century

To understand the conceptions of science at the beginning of the twentieth century, it is necessary to know the discussions around the political reform of the first Batllismo embodied in educational and rural reform, paying attention to the main actors and philosophical influences to which they resort to substantiate the actions they carry out in this context.

Barrán and Nahum argue that the first Batllismo adopted a perspective that valued scientific knowledge as a fundamental tool for progress and social development. This perspective was reflected in the promotion of policies based on empirical data and in the implementation of educational reforms that emphasized the teaching of sciences and technology.[22]

Scientific knowledge was thought to inform and support reform policies in areas such as education, public health, agriculture, and industry. For example, public education programs were introduced that included the teaching of natural sciences and agricultural techniques to improve the training of the population and promote the economic development of the country.

---

[21] Williman (1884a, b, 111).

[22] Barrán and Nahum (1988).

On the other hand, Gerardo Caetano offers a complementary view on the role of science in the era of the first Batllismo in Uruguay, also highlighting the importance of scientific knowledge as an integral part of the political and social transformations driven by Batllismo, but in his analysis, he emphasizes that Batllismo not only promoted policies based on scientific knowledge, but also sought to democratize access to it. This was reflected in the expansion of public education and in the creation of institutions dedicated to research and scientific dissemination. However, he also recognizes that this way of understanding the purpose of scientific knowledge coexisted with tensions and debates around its application and its social implications.[23]

One of the relevant actors in this period is Eduardo Acevedo, who assumed the rectorship of the University of Montevideo in 1904, just when the cycle of civil wars in our country ends. With a career that spanned roles such as professor, journalist, and politician, Acevedo had acquired a solid reputation both inside and outside of the University. Although his period at the head of the institution was not prolonged, his name is linked to one of the most significant reforms of the university structure. During those years, two important innovations emerged: The promotion regime, an educational reform that emerged in response to criticisms of the exam system used to evaluate knowledge. Secondly, the need to review the very orientation of the University. These aspects are fundamental to understand the type of training that was sought to promote among those who lead the course of science in the country. Through this initiative Acevedo sought to link higher education to one of the main sources of national wealth: the agricultural sector, which until then was scarcely developed and poorly technified. In Acevedo's reports and memoirs, in the Annals of the University of 1906, it can be read that he insists on the examples of the German seminars, comparing the results of French and German teaching, pointing out that the German student is placed from the first day in the category of experimenter:

> In contact with the realities of the world and only after he has experienced and knows these realities, he learns the theory (...) The essential in education is the development of the student's personality, the formation of habits of work and personal research. The secondary is the transmission of knowledge; instead of furnishing the head, one must form the head.[24]

Eduardo Acevedo had the intention of adapting our universities to the current orientations of European universities, especially German ones, which tended to concentrate technical functions. From this purpose arose initiatives such as the creation of the faculties of Agronomy, Veterinary Medicine and Commerce, each with administrative autonomy but under a common direction. He thought that the university cannot be disinterested in the decisive aspects of the national economy and that it must promote the economic development of the country. These proposals for university reorganization, creation of new faculties and construction of buildings required legal support for their implementation, and Rector Acevedo found the necessary economic support in the government of Batlle and Serrato. Indeed, Acevedo's

[23] Caetano (2007).

[24] Acevedo (1906, 364).

proposal for University Reform seemed to coincide with the Batllista policy to strengthen the State and promote the education of citizens for the sake of the development of national industry.

> Ministry of Development. Montevideo, September 15, 1906:
>
> Having examined the plan presented by Professor Backhaus for the organization and operation of the Agronomy Section, whose main purpose will be to train Agricultural Engineers, Agricultural Experts and Rural Foremen to provoke original scientific investigations and stimulate the development of national agronomy through tests, experiences and analysis, consultations and reports, a fundamental plan that is complemented with the foundation of the Farm or Model Estate. For the reasons set out in the report of the opinion-giving Commission that has merited the approval of the Secondary and Higher Education Council, The President of the Republic Decrees:
>
> Art. 1.° The general Plan for the organization and operation of the Agronomy section of the Faculty of Veterinary Medicine and Agronomy that is created by decree of this same date is approved.
>
> Art. 2.° The University is authorized to acquire the material for research and teaching (the installation) estimating approximately in the sum of ($60,000) sixty thousand pesos; to propose the staff corresponding administrative staff and the assistants, foremen and laborers who appear in the accompanying budget; to provide the chairs that must operate in the first year with the endowments established in that and to carry out monthly expenses inherent to the operation of the School and the Model Farm.(...)[25]

The foundation of the Faculty of Agronomy and Veterinary Medicine in 1906 with the figure of Alejandro Backhaus in the direction of the Faculty of Agronomy highlights the European influence on a new conception and practice of science that was beginning to take shape at the University of Montevideo. Where a vision of science as a practical tool is adopted, in this case, for agricultural development, focusing on the training of qualified professionals and the generation of new scientific knowledge applied to local reality.

Through the Faculties of Medicine and Mathematics, the effort to integrate physics into the Uruguayan academic sphere, a task that was influenced by the challenges of the time and the context political and social. The creation of these scientific faculties, led by figures like Jacinto de León and Claudio Williman, show that the inclusion of physics responded both to the country's needs and to a growing appreciation of the sciences in the construction of a national project. In this sense, it is clear that physics was not only valued for its practical applications in engineering, medicine or agriculture, but also for its role in the modernization of higher education and the training of specialist technicians. However, the difficulties in effectively implementing the teaching of physics, marked the development of the discipline in its early decades, mainly due to the lack of infrastructure and suitable materials and the non-existence of a solid scientific community.

This political context may have provided a strengthening of the teaching of sciences at the University, however, the markedly practical character of the scientific disciplines, as observed in the different faculties, limited to some extent the development of a research culture in the modern sense. Although significant advances

[25] Journal of the Agronomy Section of the University of Montevideo (1907, 199).

were made in technical training, the challenge of creating an independent scientific research culture remained a pending task in the following decades.

Bibliography

Acevedo, E. 1906. University teaching in 1905. *Annals of the University of the Republic*.

Ardao, A. 2008. *The University of Montevideo. Its Historical Evolution*. Montevideo: University Editions.

Barrán, J. P., and B. Nahum. 1979. *Batlle, the landowners, and the British empire. Uruguay in the early twentieth century, Vol. I*. Montevideo: Ediciones de la Banda Oriental.

Barrán, J. P., and B. Nahum. 1988. *The University Reform and Social Policy in Uruguay*. Montevideo: Eastern Band Editions.

Caetano, G. 2007. *Contemporary History of Uruguay: From the Colony to the Twenty-First Century*. Montevideo: Ediciones de la Banda Oriental.

Cahan, D. 1985. The institutional revolution in German physics, 1865–1914. *Historical Studies in the Physical Sciences 15* (2): 1–65.

De León, J. 1896. Medical electricity. *Fortnightly Review Faculty of Medicine*1–7.

De León, J. 1899. Lessons in medical electricity. *The Debates* 2 (4): 5–10.

De León, J. 1905. *Lessons in Medical Electricity*. Montevideo: Archive of the Faculty of Medicine, University of the Republic.

Duck, F. A. 2014. The origins of medical physics. *Physica Medica 30* (4): 397–402.

Herrera, F., and R. Gorlero. 1959. *History of the Faculty of Medicine, Volume I, II and III*. Montevideo.

Journal of the Agronomy Section of the University of Montevideo. 1907. Montevideo. Graphic Workshops A. Barreiro and Ramos.

La Acacia. 1873. *Weekly organ of masonic interests*. Year 1, No. 1.

Paris, B. 2009. *The University of the Republic. In the Formation of Our Liberal Conscience*. Montevideo: University Editions.

Rodríguez, W. A. M., and P. N. Martínez. 2020. Teaching of acoustics in Uruguay. Origins, facts and protagonists until 1930. *ECOS* 1 (2): 23–21.

Serrato, J. 1938. Speech. In *Bulletin of the Faculty of Engineering*. Special fiftieth anniversary issue, 12–15.

Williman, C. 1884a. Introduction to the physics course. *University Society Journal* 1:3.

Williman, C. 1884b. Applied Physics. *University Society Journal* 2:16.

Chapter 3
Physics and Its Problems at the University at the Beginning of the Twentieth Century

At the beginning of the new century in Uruguay, evidence can be found of some significant transformations in the study of Physics at Udelar. Firstly, the separation of the treatments of some physics problems, such as electricity, from those philosophical traditions that were part of the development of the discipline in the nineteenth century, is definitive. Secondly, it is not difficult to define the study of physics from a series of problems that characterized its own identity, which several professors began to take an interest in. Undoubtedly this coincides with those great milestones in the History of Physics, those related to the definition of thermodynamics, advances in knowledge of atomic structure, and of course, everything related to the theory of relativity.

This chapter highlights those scientific, political and ideological conditions that influenced the destiny of professionalization of physics in Uruguay. From general national conditions, to specific academic debates, what this section aims to provide is greater clarity to the discipline of the physical sciences once it managed to present itself away from all philosophical thought. Although this task was not the cultural characteristic of the country's great academic elite, some actors and some concrete proposals are what allowed us to begin to think about a more scientific, professional physics, updated to the world. And somewhat paradoxically, the defense of the country's first academic and scientific projects were nourished by an important philosophical reflection that was more associated with the advances of international physics.

J. A. Queijo Olano, E. Mozo Meneses, *Giants Upon Our Shoulders: A Philosophical History of Physics in Uruguay*, Studies in History and Philosophy of Science 65, https://doi.org/10.1007/978-3-031-97729-9_3

3.1 Modernisation of the University

From 1908, a series of reforms began at Udelar, which mainly promoted changes in its organization. The most relevant was that the university ceased to function exclusively through a central government, which until then was made up of the four deans of the existing faculties (Law, Medicine, Mathematics and the Section of Secondary Studies). From that date governments were established for each of these Faculties, modifying and initiating a federalist logic of organization. This form of organization preached, incipiently, the possibility of including, in the mission of the university, the function of scientific research as a legitimate matter of universities. This was a point of tension in parliamentary circles as well as in university ones.[1] Another relevant aspect in the institutional organization of the university was the inclusion of the participation of students in the areas of government of the institution. This aspect will also be fundamental in the discussion of the development of scientific research, from the Reform of Córdoba, a student movement that spread throughout the continent.[2] In a 1928 text, which recapitulates the importance of this event in the context of the transformations of the public Latin American universities, with a focus on Uruguay, the Uruguayan journalist and writer Carlos Quijano (1900-1984) wrote about the task of transforming the model of a *professionalist* university, that is, one that only cares about training liberal professionals

> *Professionalist University.* It is the essential feature of our teaching, and is, therefore, the one that the Reform fights most. Without doubt society needs professionals and the State has the duty to control the preparation and action of these. Therefore, the University can never stop having professional training schools. But does the university institute fulfill its mission by reducing it to fulfill fully its mission, to this task? No. The University must also be a great center of culture, a great center of non-professional scientific research, if we may use the term.[3]

These transformations in the country's largest house of studies, were in line with the government's interest under the presidency of Batlle y Ordoñez, to achieve modernization in all spheres of the State.[4] This idea came to play an active role in the country's economy through the promotion of industrialization processes, the implementation of fiscal policies and the special participation in the agricultural, social policies and labor legislation fields. In terms of education, the most relevant was the process of expansion of secondary education to the interior of the country, as until then it was concentrated in the capital. Some of the new faculties emerged in the Batllista period, such as the mentioned separation of the Faculty of Mathematics and Related Branches, consolidating the autonomy of formations oriented towards productivity and national industrialization, such as architecture and engineering.

---

[1] Oddone and Paris de Oddone (2010).

[2] Finocchiaro (2013).

[3] Quijano (1941).

[4] Yaffé (2000).

This impulse found both strong adhesions and strong resistance, led mainly—as could be expected—by the more traditional classes, which, in turn, managed to involve the rural movement, which did not share the visions of the capital. Thus, the reformist impulses of the capital and the traditionalist caution of the interior of the country were political expressions that alternated in the government through cycles that show the way in which Uruguayan democracy was expressed. But what is relevant, for the case of physics, is to note that these alternations and visions about modernization processes were also part of the problems that ran through the university halls. More directly, the discussion about the relevance of scientific research, about the real insertion of science in university curricula, at the beginning of the 20th century, is part of a broader debate, which encompasses the economic and industrial destiny of the country.

At the beginning of the new century in Uruguay, there can be evidence of some significant transformations in the studies of Physics at the Udelar. First, the separation of the treatments of some problems of Physics, such as that of electricity or thermodynamics, from those philosophical traditions that were part of the development of the discipline in the nineteenth century, is definitive. Second, it is not difficult to define the study of Physics from a series of problems that were characterizing an identity of its own, on which several professors began to take an interest. Undoubtedly this coincides with major milestones in the history of Physics, those related to the definition of thermodynamics, advances in knowledge of the atomic structure, and of course, everything related to the theory of relativity. All these advances were not only known, it could be said that they influenced the way in which Physics achieves autonomy as a field of knowledge in Uruguay, a process that is crowned in 1941 with the creation of the Institute of Physics in the Faculty of Engineering and Related Branches. These decades that go from 1900 to 1940, allow us to distinguish some areas that will mark the path of Physics in Uruguay for the twentieth century. They are: electricity, radiation, thermodynamics and relativity. In a very general way, these are many of the concerns that we see reflected in physics programs and publications of the time. But it is also fair to say that this problematic definition did not yet manage to find a space for appropriate institutional development at the Udelar. Contrary to nearby experiences in the region, such as those of the Universities of La Plata (UNLP), Universidad del Litoral or Tucumán (Argentina), or that of San Paulo (Brazil), in the early decades of the twentieth century the Udelar did not manage to find the way to provide adequate spaces for the development of Physics in a professional manner.

Although the University had left behind the philosophical discussions that centered the discussions in its Board of Directors, the problems of the new century focused on being able to guide new directions to the type of education that the institution provided. The difficulties for graduation, the need or not for a final exam in secondary education (which was in charge of the institution since its foundation), the relevance of a training aimed at obtaining degrees, all these discussions accompanied the early years of 1900. It is also pertinent to highlight that, in these decades between centuries, the university budget was strengthened compared to the meager situation that the institution lived through almost all of its existence in the nineteenth century. This budgetary commitment directly impacted the incipient scientific aspirations of the university class more inclined to positivist ideas, which found in Rector Alfredo Vázquez Acevedo their great reference. As an example, in 1884, the military government of Máximo Santos allocates a large sum for the installation of physics cabinets, geography, cosmography, equipment for a chemistry laboratory, as well as natural history book collections, and scientific bibliography.

Let's remember that the University had started, very modestly, its path towards the search for the development of careers in more scientific areas when in 1873 and in 1888 it creates its Faculties of Medicine and Mathematics. Although the first had a clear profile that was especially demanded in Montevidean society long before the foundation of the Faculty, in the case of the Faculty of Mathematics it was quite different. Within the profiles developed in that institution three professions of great importance could be clearly highlighted: that of surveying engineer, architect and engineer. It is important to highlight this for two aspects: the Faculty of Mathematics was, then, the faculty that gave rise to two future Faculties: the of Architecture and the Faculty of Engineering and Related Branches (1915). On the other hand, scientific training was always subsidiary to these professions: the physics and mathematics developed there had the purpose of a specific application. Only when the Faculty of Engineering manages to establish itself as an autonomous institution, and when the physics achieves an exponential development in the world in the first decades of the twentieth century, only from that moment can we begin to trace a physics that defines its own problems, trying to follow the developments that are taking place in the world.

The training profile that had characterized the University, that oriented towards a ruling political class, began to be questioned by a few university actors. The need was manifested to train a new type of professional, that specialized in the challenges that industrialization presented. This new graduate, although not the prototype of the *scientist* that we can recognize today, began to show interest in problems inherent to science. In the case of physics, many were the engineers who as they fulfilled roles in activities of development of civil and industrial engineering of the country, dedicated time and reflection to problems like the novel theory of relativity, advances in relation to the discovery of the atomic composition of matter, and other strictly physical problems. The incipient process of industrialization that the country experienced allowed the ruling classes university to rethink the way in which new students could contribute to the task of national construction, from new technical knowledge associated with new challenges. However, this vision that sought to promote "theoretical" science within the University, found strong resistances. Characters like Bernardo Kayel and Eduardo García de Zúñiga were protagonists of a foundational episode-not for its success but for its original proposal-, within the Faculty of Engineering: the creation in 1916 of the Higher Institute of Physics. This proposal was framed in a substantive change that both encouraged in the Faculty: to attend to the growing needs for specialization of knowledge in Engineering, and to think of the University as a space for development and defense of a "high culture" where science be your bastion.

3.2 Specialization and National Timing

Prior to the most precise examination of the scientific development of each of the areas of Physics that, at the beginning of the century, manage to arouse the interest of not a few university students, it is good to show the state of discussion about the relevance of science in those decades, within the University. In particular, we are interested in recognizing the conception underlying a series of dichotomies that were manifested in university discussions: encyclopedism or specialization; applied or disinterested studies; theoretical or applied science.

In 1912, Professor Octavio Hansen proposed to expand preparatory studies in Mathematics so that students entering the Faculty of Engineering would have better preparation. Octavio Hansen was a professor of Physics and a member of the Board of Directors from 1909 to 1913, when the reorganization of the University promoted by the new Organic Law of 1908[5] was discussed. Hansen advocated for special training in Mathematics prior to entering the career, which would later be the basis for the changes that the Council would approve in 1914. Review and expansion of elementary mathematics, introduction to higher mathematics, inclusion of higher algebra and analytic geometry were his proposals to strengthen the theoretical training of new students. His defense of the proposal recognized that mathematics was a special discipline within the set of knowledge, and for that reason they should be treated differently:

> It can be assured that if there are studies that need special preparation before starting them, they are undoubtedly those that are developed in it [Mathematics]. No one ignores that Mathematics constitute the basis of them and therefore, the study of application subjects should not be done until their perfect knowledge is possessed. However, with the current system, it has not been possible to avoid the simultaneous study of practical subjects and pure mathematics.
>
> This confusion of studies of different nature, results in, not only insecurity due to the lack, in many cases, of indispensable mathematical knowledge for the proper understanding of certain subjects, but also makes the student neglect the application subjects, trusting that their exam lends itself more to come out successful with answers not entirely precise, while he dedicates his greatest attention to mathematics because he knows that these do not admit tricks and vagueness, and that it is necessary for him to reach the solution of the problem posed, going straight and relying on logical reasoning.[6]

His proposal was rejected and underwent significant modifications. On the one hand, there was an excess of content in the preparatory training of students aspiring to enter the University, which would make entry even more difficult than it already was. But the greatest resistance came, precisely, from the argument of some counselors about the lack of general knowledge that would imply for future engineers the absence of subjects such as History, Philosophy and Literature. In this sense, Carlos Vaz Ferreira argued, who proposed to include some of these subjects in the

[5] Renaissance (1926).

[6] Annals of the University (1912, 39–40).

reformulation that was being proposed. In a way, the proposal to emphasize theoretical studies in mathematics for students who were going to enter Engineering, found a resistance that-as we will see-will be constitutive of this whole period in relation to generating more specialized studies in areas of theoretical sciences. This resistance adopts two major strategies, which are always seen as antagonistic in the contexts of budgetary restriction of the University, where the meager resources must be oriented in one direction or another, but never in two or several: first, the uselessness of theoretical training for professional performance; and second, the defense of a "culturalist" training, which demands the existence of a broad training for professional performance (almost always matrixed in humanities content).

Thus, Octavio Hansen's proposal did not achieve its goal but left a mark that would continue 2 years later with Eduardo García de Zúñiga. Octavio Hansen should be seen as one of the first representatives of that small group that, between 1910 and 1930, promotes the idea of generating changes in the training of engineers, especially in the sense of providing a deeper training in the areas of mathematics and physics.[7]

In the Faculty of Engineering, between 1914 and 1916, a series of disputes took place that found some professors of the Faculty, such as Eduardo García de Zúñiga and Bernardo Kayel, in opposition to the dean Juan Monteverde. These discussions manifest the latent tension of what could be considered the central function of the University, in the aforementioned dichotomies, and in the real possibilities of the institution.

In 1914 Eduardo García de Zúñiga presents a change in the curriculum of the Engineering career that implies a complete reformulation of the teaching of mathematics. The central purpose was to make the mathematical studies that were taught for engineers, something more than basic tools for the practice of the trade. Although the discussion focused on a significantly smaller change, such as the proposal to replace Projective Geometry with Analytic Geometry among the subjects of the career; it is undeniable that the central purposes of García de Zúñiga were to cultivate the development of the scientific discipline for the sake of proper mathematical knowledge. If this episode seems relevant in the History of Physics at the University, it is because it allows to shed light on how it was possible the development of an early theoretical physics in the context university of the early twentieth century. To do this, we must deeply understand the changes proposed by García de Zúñiga as well as the resistance he encountered from Dean Juan Monteverde.

In 1914, the proposal for *Studies on preparatory and higher mathematics* was published by García de Zúñiga and published by the Journal of the Polytechnic Association of Uruguay, which had been presented to the Council and approved a year later, after an intense discussion. In the statement of reasons, the author argues that at the time he is developing his proposal, mathematics is undergoing a significant change in its methodological conception, which makes it necessary to reflect

[7] It will also be Octavio Hansen who will suggest in 1923 to the young Enrique Loedel Palumbo (whom we will deal with later) that, to professionally dedicate himself to studies in Physics, the most convenient was to travel to train at the Physics Institute of the University of La Plata.

this transformation in the university curriculum. There are two ways to understand the role of mathematics for García de Zúñiga: on the one hand, a cultural and educational purpose; on the other, a utilitarian purpose. And these purposes, which will find their correlation in the preparatory and higher mathematical training, must be given when it is possible to clearly define what relevance mathematics has in the cultural and professional formation of the individual.

> The first question we must examine, albeit very briefly, before determining the subject matter and extent of our programs is, therefore, the following: What role do Mathematics play in general culture, whether considered in themselves, or considered in their applications to other pure sciences? As an independent discipline, Mathematics, along with Philosophy, contribute primarily to the development of the highest qualities of intelligence. (…)
>
> To this general and latent utility of a solid mathematical preparation, is added the concrete advantage of acquiring truths and concepts applicable to all other sciences and indispensable for some of them.
>
> Today, mathematical science is much less important for the knowledge-very real and precious however-that it directly composes, than for the fact that it constitutes the most powerful tool that the human spirit can employ in the investigation of natural phenomena». (Comte, *Cours de Philosophie positive,* second lesson).
>
> Astronomy and, to a large extent, Physics, already clearly have the character of true mathematical sciences; and, if Chemistry is still waiting for the Newton that Emilio du Bois-Reymond announced commenting on Kant's opinions on natural sciences, its evolution, in the sense of becoming an exact science, is more apparent every day.[8]

Two relevant aspects of how the beginning of mathematics is conceived in García de Zúñiga's proposal. First, taking as a reference a hierarchy and ordering of the sciences in a positivist way, mathematics and philosophy are the disciplines that are identified with the development of intelligence. This idea is not new in general scientific thought nor in the Uruguayan context. In 1886, the surveyor Nicolás N. Piaggio wrote in the Bulletin of the Society of Sciences and Arts, an article titled *Parallel between Philosophy and Mathematics*, trying to provide his position on a discussion that was taking place in that Society about which discipline was superior to the other.

Piaggio, a renowned agronomist who was a professor of cosmography at the University until his death in 1918, is perhaps one of the most important responsible for the cultivation of calculus, algebra and the use of the rudiments of mathematics for applications in the field of agronomy. But without a doubt his thinking about pure mathematics, its importance in the formation of the individual, does not escape the idea of the time, still dominated by the prevailing conception of philosophy as the mother discipline of all others. In this sense, the relevance of mathematics is never justified by itself, but in the gear that only philosophy can make sense.

> Erase the morality of the peoples and leave all the individuals that form them, to mathematical calculations, and then you will have a semi-brutal people that would only be guided by

[8] García de Zúñiga (1914, 5).

> the properties of quantity and by the depraved customs that immorality brings with it. But suppose now, that a people is moral and not mathematical, and you will have a happy people, who without knowing what material greatness is will at least know how to live as befits the cultured man in the sublime sphere of morality. (…)
>
> It is known by all that the study of mathematics is never followed with as much ardor as that of philosophy: I speak in general. And this, why? Because the study of pure mathematics, the first to be carried out, is somewhat antipathetic in its principle; it is difficult to dress it up with some attraction that erases in the student's mind that bad idea that he formed of it.
>
> Does the same thing happen in philosophy? No: from the beginning it is already awakening in the student's mind the desire to finish it.[9]

If it is important to show this tradition, still in force in the first decades of the twentieth century, it is because the change in García de Zúñiga's proposal not only took place at the level of the mathematics curriculum itself; it implied a change of perspective that gave autonomous value to the study of sciences in their non-applied and pure sense. The tradition that resisted the changes in the mathematics plans combined two strong positions that were dragged from the nineteenth century: on the one hand, the reign of philosophy as an expression of the moral and cultural thought of society, which should spill over onto the other sciences; on the other hand, the desire not to contaminate the formation of professionals with knowledge of little practical use for the development of their technical training. Thus, this resistance was led by Dean Juan Monteverde, in the various disputes that crossed the Faculty of Engineering in that second decade of the twentieth century.

Monteverde made public his discontent with the decision of the Council of the Faculty of Engineering, and published his discharges against the plan of García de Zúñiga a year after the new plans were approved. It was not only about a specific disagreement, as we will see, but about the opposition to the emerging expression that advocated for an autonomous development in mathematics and physics within that house of studies:

> the two trends that are in conflict in our Faculty, so that they give their vote in the future elections to candidates who respond in the first term to the well-understood interests of the Faculty, which cannot be other than those of its teaching, which is the basis of the credit of the professionals it forms.[10]

The "trend" defended by Monteverde sought to stop what was considered "an unnecessary overload of theoretical studies" promoted by the new mathematics plans led by García de Zúñiga. Monteverde understood that these changes would be carried out "at the expense of the exercises and practices that are indispensable in the professional studies of engineers, architects and surveyors."[11]

The underlying argument that Monteverde holds is not in the intention *per se* to raise the knowledge of pure mathematics of students, as García de Zúñiga intends, but in that the latter does not take into account the real conditions of the Faculty at

---

[9] Piaggio (1877, 189–190).

[10] Monteverde (1915, 17–18).

[11] Monteverde (1915, 18).

the time of making his proposal. A sort of pragmatic thinking is what moves Monteverde to reject this initiative, which will be inconclusive according to his opinion, given the very low level in mathematics with which the students come from preparatory studies.

> Our Faculty of Mathematics is in formation, and although we have managed to equip it with valuable material elements, teaching aids, it is necessary to form the stable teaching staff, selecting it among the most capable and studious technicians who are in a position to dedicate to their improvement and to the classes the necessary time.[12]

3.3 Geometrical Debates

In the end-of-year sessions of 1914, we see that one of the most relevant aspects of the new mathematics program promoted by García de Zúñiga has to do with transforming the studies of Geometries in the teaching of engineers. In the session of November 14, García de Zúñiga proposes to defend the proposal of changes to the Study Plan for mathematical training in the Engineering career. Although this proposal had been approved in previous sessions by the FIA Council, Dean Monteverde, absent in those sessions, asked for the reconsideration of what was approved. Specifically, the discussion focused on the proposal to eliminate Projective Geometry from the Engineering career, and replace this training with Analytical Geometry, moving the former to secondary studies. Arguing mainly from what was done in other universities in the world, such as Rome, the University of Buenos Aires, and many others, Dean Monteverde strongly opposes García de Zúñiga's proposal.

In response, in the session of the mentioned date, García de Zúñiga presents elements that become interesting to analyze not so much for the way of understanding his attempt to introduce pure mathematics disciplines, but because that same attempt-and its application in the years following-, can be understood as the most evident germ of the development of a basic science until then recognized in the University. And if this hypothesis can be plausible, we can say that the challenge of introducing new geometries (oriented mainly towards mathematics) in the training of engineers, will make it possible for new engineers to approach the challenges of the new physics.

The question, then, is to understand why such a large conflict develops between these two geometry courses. As we saw, it is not so much about the relevance or not of the contents of them, what the diatribe is about, but to consider what place they would occupy in the general training of engineers. And here it is worth noting what implies the training of projective and analytical geometries. As can be guessed, the first one has to do-historically-with the projection of a figure or body from one space to another, to the point that it was particularly developed by artists like Brunelleschi or Alberdi. In this sense, projective geometry, although it follows some

[12] Monteverde (1915, 19).

principles or rules that we could call "mathematical"—a straight line will always be projected straight or that parallels will be projected existing a point of intersection between both (vanishing point)-, does not lose its relationship with geometry. And it seems natural that the training of engineers should handle these elements. Analytical geometry, as such, does not differ substantially from the projective one. Its initial interest has been the calculation of conic sections, and only recently from modernity (Descartes and Fermat), it acquires a complexity with the use of equation calculation introduced by both in a parallel and almost simultaneous way, showing that they can be *coordinated* the points of a curve with a type of function algebraic, if we postulate different degrees of equations. While it is not about a geometry of "higher" abstraction (authors Stilwell today They refer to analytic geometry as the "elemental residue" of what algebraic geometry and differential geometry leave behind), it was the step that García de Zúñiga intended towards these contemporary developments.[13]

> From a logical point of view, Projective Geometry not only precedes higher mathematics, but also precedes Elementary Geometry itself. Metric Geometry, says Couturat, is historically prior to Projective Geometry; it is the Elementary Geometry that everyone knows and that all geometers have practiced from Euclid to the nineteenth century. However, it is logically posterior to Projective Geometry. And Bertrand Russell arrives at exactly the same conclusion after a long and deep analysis. Pedagogically, Projective Geometry, due to the simplicity and the short number of its first notions, is within the reach of beginners, without anything preventing it from being taught either before or at least immediately after Elementary Geometry.
>
> (...) Projective Geometry has no direct applications to technical sciences, such as infinitesimal calculus, rational mechanics, analytic geometry. The eminent Italian professor Colombo, Director of the Polytechnic School of Milan, a great supporter of the high mathematical culture of engineers, said, however, in a recent speech: It cannot be pretended nor would it be necessary, to give young engineers a mathematical culture equal to that which the Faculty imparts to its students; it will suffice, on the contrary, to suppress or reduce to a minimum some courses that only have importance for mathematical culture, and intensify those of evident necessity, that is, Differential and Integral Calculus, Analytic Geometry and Rational Mechanics, including in these subjects, by way of introduction, the few chapters of complementary Algebra necessary for the resolution of equations and the notions of Projective Geometry eventually required for the course of Descriptive Geometry and for the use of Graphic Statics in the courses of Mechanics and Construction.[14]

In summary, Dean Monteverde was extremely opposed to enabling the theoretical study of branches such as mathematics and physics within the Faculty of Engineering. The main argument, his tenacious opposition, rested on the double fact that: the main task of the Faculty was the training of professionals, and in this sense the contents should be adjusted to this objective; and it was also materially impossible to develop any type of scientific initiative either in the area of mathematics or, as we will see, in the area of physics.

---

[13] Stillwell and Stillwell (1989, 121).

[14] Council Minutes (1914).

3.4 First Proposal for the Creation of an Institute of Physics (1916)

This type of discussion was frequent throughout almost all of 1916 in the Council of the Faculty of Engineering. A commission, composed of Kayel and García de Zúñiga, proposed to the Faculty a series of transformations that were summarized in three major actions: one, the total reform of the curriculum; two, the creation of the Higher Institute of Physics, and three, the annexation of the Institutes of Geology and Drilling, and Industrial Chemistry, which were in the Ministerial orbit.

While the most interesting for the purpose of this study is to determine the nature of this first national and university proposal for the creation of a Higher Institute of Physics, it should not be understood in isolation. The first point in question had to do with a general change that the commission, led by Kayel in the company of García de Zúñiga, proposed to the Faculty Council regarding the study plan. The proposal began by modifying the title of Bridge and Road Engineer to Civil Engineer. This change had a reason: the training that was obtained (or should be obtained) when studying the engineering career, was so general that it should be reflected in the title to be awarded. But this simple fact, for Kayel, opened the opportunity to think about the development of specialized formations, not contemplated in civil engineering *stricto sensu.*

The proposal was specifically to introduce specializations in *electro-mechanics* and *industrial chemistry,* and this was the first point of divergence that both professors had with the then Dean Juan Monteverde. The latter argued that:

> (...) it is a mistake to create permanent studies of well-defined specialties in national engineering, because it is out of place; because providing them with a location, teaching material, faculty, etc., would mean a disproportionately huge expense in relation to the utility it would bring to the country and the student youth. This issue of specialization was discussed 5 years ago by a Commission of professors from the Central School of Arts and Manufactures in Paris, and the conclusions were against the establishment of specializations in the School. It was recognized that rather than being beneficial for the engineers graduating from the School, they were harmful. In Buenos Aires, the same issue was also discussed on the occasion of the suppression of the teaching corresponding to the specialty of mechanical engineer. The Commission appointed by the National Center of Engineers reported in favor of suppressing this specialty, since the existing mechanical engineers could not find a position in the country. Now, if in France and Argentina, countries of greater industrial importance than ours, it has been declared by competent professors, by engineers with long professional practice, that it is not convenient to specialize the studies of engineer, what will we say here, in a country that has no more industries than the most elementary and simplest, which barely give rise to the occupation of half a dozen engineers, who in reality would not need more knowledge than that of a civil engineer. What What would we do with the mechanical and industrial engineers that make up our Faculty? Is it possible to suppose that more than a dozen of them can be employed in public jobs?[15]

[15] Council Minutes (1916, n/d).

This will be the dean's main argument in all the months in which this controversy will persist about the possibility of developing more advanced knowledge in the branches of engineering, as well as installing scientific research spaces in the university: there are no national conditions to host professionals of such specific specializations; the cost of embarking on such an enterprise would only be profitable for a few students and, in such a case, a safer and less costly strategy would be to send them to train abroad. And again, as in the discussion about geometries, the arguments about what was done in other latitudes served as a reference to guide what type of teaching should be established in the Faculty.

Kayel does not agree with these arguments. The country can and should insert technical professionals in areas where it still has no specialists, which would ensure job opportunities for the few engineers who would want to develop in these areas. Monteverde understands that this is a proposal "out of reality", because simply training specialized engineers for the industry will not make the country more industrial. Uruguay at that time is, for Monteverde, in a "primitive industrial period". By April 24, after a long discussion in the two previous Councils, the session understands that a decision must be made, and the members of the Council seem to agree to approve the suggestions prepared by the Commission. But Dean Monteverde insists. In addition to his more conceptual arguments, the practical reason, the costs and difficulties of incorporating these specialties, makes the proposal even more difficult. A majority of Council members are in favor of accelerating the discussion to achieve approval of the changes suggested by Professor Kayel, but they find in Dean Monteverde a significant obstacle:

> I do not believe that the Council will face the responsibility of wanting to carry forward the purpose expressed by the majority of its members, to accelerate the discussion in which we are engaged, to arrive as soon as possible at the approval of the project proposed by the reporting commission; such haste in dealing without all the required study in matters of such importance as those involved in Mr. Kayel's project, would be more than a disregard for the members who believe that project to be harmful; it would be failing in our duty to look after the well-understood interests of the Faculty.[16]

Thus, in that session, the dean presents to the members of the Council, a proposal for reorganization of study plans, which he drafted (during Holy Week, he notes), based on studies of Schools of Arts and Manufactures in several European cities, and which was the first obstacle to stop the transformations promoted by Kayel. Monteverde thus begins to sway the vote of the vast majority of the Council. Eng. De Mula suggests studying the new proposal and moving on to the initiative to create the Higher Institute of Physics.

[16] Council Minutes (1916, n/d).

3.5 Physics, a Luxurious Subject

In the session of May 2, Kayel begins a defense of the proposal to create a Higher Institute of Physics in the Faculty of Engineering and Related Branches, accompanied by García de Zúñiga. It should be remembered that by then, the Faculty of Engineering already had two Institutes: the Materials Testing Institute and the Machines Institute. A question that arises in reconstructing this history is to understand the reasons that led an engineer like Bernardo Kayel, to take on the impulse to propose the creation of a Higher Institute of Physics.

Bernardo Kayel, a native of the city of Paysandú, manages to graduate in 1901 as a Bridge and Road Engineer. A few years earlier he had been a professor of Mathematics in the Faculty of Mathematics. During his university years he begins to get involved in what will be his main area of professional development: radiotelegraphy or also called wireless telegraphy.

Inspired by the advances made by the German physicist Heinrich Rudolph Hertz, Bernardo Kayel's training was as formal as it was self-taught.[17] By the time Kayel was an engineering student, electromagnetism had found its solid foundations, from Maxwell's equations, in dielectric currents, but a work by Hertz in 1888 took the problem of electromagnetic waves at a distance to a new level of advancement: he demonstrated the propagation of electric and magnetic forces in the air, using Maxwell's contiguity law. Hertz seemed to have settled not only the old dispute about the nature of light, demonstrating its wave essence and empirically verifying the electromagnetic theses, but above all he managed to show the unifying power of mathematization in electric forces, using Maxwell's theory, which would later give rise to a theory of radiation.

If we stop at this central aspect of the history of electrodynamics studies, it is for a double reason: Kayel, as we saw, proposed for the Faculty of Engineering and Related Branches a specialization in *electro-mechanics*, following the recent trends of the discoveries that were made in Europe in the days when he was a student. The second aspect has to do with the fact that this appreciation for advances and new discoveries in Physics were for Kayel a clear guide to the practical role that such advances could to offer to developing nations like Uruguay. These are scientific advances that Kayel studied relatively shortly after they were discovered, when he was still a young man in training, also highlighting the practical and applicable aspects of Hertzian waves. In fact, this last aspect was very closely linked to the professional future of Kayel himself, who would take on a role as president of the Administration of Electric Power Plants and State Telephones (U.T.E), in the de facto government of Gabriel Terra, and then continue his work on several significant works in Uruguay, such as the Rincón del Bonete dam.[18] It does not seem very

[17] Thomasset Trakalo (2019).

[18] Waiter (2019).

strange to suppose that the vision behind the project to create a Higher Institute of Physics is for Kayel, ultimately, a more specialized form of engineering in the concrete application of new advances.

In the discussion given on May 2, 1916, Kayel begins arguing about the relevance of creating a Higher Institute of Physics, highlighting in the first instance the way in which it could be financed, anticipating the problem that had already been presented by Dean Monteverde. According to Kayel, the financing could be produced either by the annexation of the already existing Institutes of Geology and Drilling and Industrial Chemistry; or by one of the tasks that the Institute would assume, that of being a space for regulation of measures, lacking in the country, where to guarantee the standards of international measures. Since 1908, Uruguay was a member of the International Meter Convention, an international treaty that had been signed in Paris in 1875. For Kayel, it would be the Higher Institute of Physics the indicated space to guarantee the country the execution international of the criteria established in such convention.

But beyond this, Kayel sought to make clear that the Higher Physics Institute would not be a center of exclusively theoretical research. It could even be said that, in view of the documentation obtained and the discussion processed in the Faculty Council of Engineering, it did not seem very clear how a research in theoretical physics, in those early years of the beginning of century. In fact, for Kayel the purpose of developing a pure science has to do with learning new scientific theories, their experimental replica so that it is a source of training. Nothing appears, in his words, of the relative to the search for new knowledge, or development of research:

> On the other hand, the Institute will not be exclusively pure science, experimental science will be cultivated with careful consideration, lending itself, in this way, to intensify the practical teaching of the Faculty.
>
> And here it is opportune to specify, to avoid a regrettable confusion, the difference that exists between the practice of teaching and the practice professional. The first consists of the exercises and experiments necessary to better assimilate scientific theories, and corresponds, consequently, to the establishments of instruction, while the other can only be the experience that gives the exercise of a profession and belongs, therefore, to extra-university activities. To the scientific practice of our Faculty, therefore, the projected Physics Institute must provide indispensable service.[19]

And to close his intervention, Kayel notes the intangible value of his proposal, the appreciation for the high culture that this type of institutions can produce. And in this sense, the Faculty of Engineering could be seen as an institution, even in comparison with other Faculties of the university itself, as a space lagging in its contribution to the intellectual construction of the nation.

> But above all these secondary services, the Institute will provide the most important service to which its creation corresponds, which is that of promoting high culture. Certain scientific investigations are logically destined for it. Because the Higher Physics Institute was not founded in a timely manner, the government entrusted the doctors with an Institute of Radiology. Doctors are only suitable for radiotherapy, radiology generally corresponds to a

[19] Council Minutes (1916, n/d).

> Faculty of Sciences like ours, and especially a Physics Institute like the one we have projected. We also need to put at the disposal of teachers and students an Institute where to repeat interesting experiences and undertake the investigations that are suggested to them by the readings of books and the dissertations of the classrooms. We have the duty to prevent some talented students from being wasted for that kind of studies. The Institutes must therefore be a complement to regular teaching, without prejudice to the practices necessary for the better assimilation of that teaching.[20]

Dean Monteverde then intervenes in the matter with a response that reflects one of the most characteristic tensions of the University: "before high culture is the advancement of our lower technical institution."[21] This will be the spirit that will mark the Dean's refusal to the proposal of Kayel and García de Zúñiga, which entails a matter of crucial relevance if we think about the mission of Latin American universities: Can we think of dedicate resources to pure research when the basic conditions of university teaching are not covered? Where should budgetarily prioritize, in science that few will follow or in increasing the technical training of many? For Monteverde, perhaps influenced by his institutional position (it is worth remembering that Kayel occupied his position in the Council as a teacher representative elected by the students), the answer is emphatically negative.

When the discussion is resumed on May 17, the dean reaffirms being persuaded that the creation of a Higher Physics Institute is a real "luxury in our environment"[22] because it lacks the required preparation.

[20] Council Minutes (1916, n/d).

[21] Council Minutes (1916, n/d).

[22] It is interesting to show the relevance that was acquired, in the periods of government of Batllismo, the discussion about where to acquire specialized and socially relevant knowledge. Juan Monteverde defends the idea that, given the small number of students interested in developing studies in areas of advanced theoretical development in engineering or physics, it was more profitable for any State to send these people to train abroad than to invest in infrastructure for national development. Ten years before the discussion about the creation of the Higher Institute of Physics. Juan Monteverde and Eduardo García de Zúñiga starred in a similar discussion relative to the idea of a construction of the engineering profession. In 1907, the new Ministry of Transport and Public Works that Uruguay had, chaired by the engineer Juan Pedro Lamolle-first minister of Uruguay in that portfolio-, decides to hire foreign engineers for the execution of certain public works that were undertaken at that time in the country. The Association of Engineers and Architects of Uruguay, whose president was García de Zúñiga, published, in its official magazine, a response to such an initiative emphasizing that the Association: "Also does not see with displeasure, despite recent failures, and being convinced that the most effective procedure would be to send technicians to specialize abroad, the hiring of distinguished specialists when there is a real need for them for the conception of public works of extraordinarily difficult nature; but does not judge it opportune, «declares it with loyal frankness», which does not exclude in any way the sincere respect to the person to whom it is addressed, twenty years after our school of engineers and architects was created, the importation to carry out relatively simple and current works, of foreigners whose exceptional competence cannot in truth be guaranteed by the small assignment that is offered to them" (García de Zúñiga 1907, 85).

In the same issue, Juan Monteverde expressed the same rejection of the Association. The works that the Ministry had entrusted to foreign technicians seemed to be usual practice among national engineers, which created an atmosphere of annoyance among the Uruguayan professional class: "Let foreign specialists be welcome if they come to collaborate with us in the important works that

> It would be more beneficial to use the money that the Institute will require for the progress of our practical teaching, aiming at the improvement and increase of teaching staff, and in particular, that of the class assistants. In addition, the state could always help those people who showed special attributes for theoretical studies, facilitating their stay in the great intellectual centers of abroad.[23]

After Monteverde's intervention, Bernardo Kayel closed the oratory. The vote was unfavorable and it will take almost 25 years for the Faculty of Engineering to finally have its Physics Institute, which will only be created in 1941. Almost as in a message to posterity, the engineer promoter of the initiative said:

> We must do science and research as far as we can; it is about an unavoidable social contribution. Perhaps we will also come to occupy a good position as an intelligent country, if not as a producer. Our wealth and strength will not make us stand out among other countries and only will we be able to impose ourselves with our culture. The institute aims to provide more horizon to scholars, providing them with the means to acquire intense culture. The country needs men of science, and the least service that the Institute can provide will be to reveal to us the most suitable, and also gather many data of great scientific value, now scattered, and that are lost.[24]

Despite seeming a message clearly affirming the need to promote science, attend within university studies to cultivation of scientific training, a doubt always arises when reading these claims: what are we talking about when we defend this idea of science, what contents-thinks Kayel-are those that are considered necessary to develop the studies of physical sciences. Some lines are what engineer Kayel seems to have in mind: an institute that can serve as a national center for the establishment of measures, where precision instruments can be calibrated and repaired; an institute that will be both theoretical and experimental physics; an institute where-perhaps-theoretical developments in X-rays and their properties can be made, in view of the fact that radiology was already a discipline co-opted by medicine (as will be seen in the next section). In summary, it is not very clear what was sought specifically with the creation of this Institute and this unclear destination of research in physical sciences does not seem to be so much a personal aspect of Kayel as a difficulty related to the type of professional university that was Udelar.

Other events, related to the transformations that occur in the first decades of the twentieth century within physics, seem to be more relevant to explain the autonomous development of the discipline in this time. An unavoidable point of these transformations is the appearance of Einstein's special and general theory of relativity, which become more important when the German physicist visits Montevideo in 1925.

---

the country needs for its enlargement; but let's not go to say abroad that to sanitize small cities, with or without biological purification, imported engineers are needed, nullifying the engineers of the country or making them play passive roles to the detriment of their dignity" (Monteverde 1907, 85).

[23] Council Minutes (1916).

[24] Council Minutes (1916 n/d).

3.6 Radiology in Dispute

The first episode of the discovery of X-rays occurred on November 8, 1895 in Roentgen's laboratory, in Würzburg, Germany, but the publication of the results were not known until early January 1896. The first human radiograph he obtained using his wife's hand and from there began a accelerated diffusion of the method, which motivated that in that first year of 1896 more than a thousand publications related to X-rays were made, renamed as Roentgen Rays. The method gained international publicity through the press, and photographers, physicists, doctors, and even merchants exploited the curiosity. This discovery was known in Uruguay a few weeks later, on March 17, 1896. The newspaper *La Razón* from Montevideo reported on experiences of X-rays that were being carried out in the physics cabinet of the University, using tubes of *Crookes*[25] made for experiences with cathode rays.

> The following photographs have been obtained: the image of a frog in which most of the details of its skeleton are observed. A turtle whose interior, despite the shell, appears perfectly demarcated the bones of the skull are all outlined. A box of leather and wood completely closed and inside which had written "Glory to Roentgen" with lead letters... A compass box, lined with velvet with a thickness of one centimeter and even covered by opaque bodies produced in the photograph very bright images of the bodies contained inside with the wonderful particularity that a pencil that was mixed with the compasses clearly revealed in the plate the graphite bar.[26]

A whole series of photographs were published in the book *Elements of Physics* in 1898 by engineer Tebaldo Ricaldoni, Uruguayan professor of Physics at the UNLP, as the first radiographs obtained in the Río de la Plata.[27] The Uruguayan magazine also reproduced them *El Estudiante*, from September 5, 1898 under the title *Photographic revolutions—The photography of the invisible—Notable experiences in our University*, magazine, rescued by Mañé Garzón. This experience was carried out by the lawyer and physics professor of the Faculty of Preparatory Studies, Dr. Claudio Williman, future rector of the University and future president of the Republic. Dr. Jaime Oliver, then a chemistry professor and future professor of surgical pathology at the Faculty of Medicine, as well as the student Ángel Maggiolo also participated.

National medicine immediately appreciated the utility of X-rays and these experiences, carried out as demonstrations and without medical utility, were repeated throughout the year. José Claudio Williman, son of Dr. Williman, dates in the biography he made about his father the moment when the first request for a radiological study of Uruguayan medicine, with clinical data included, was made:

[25] This instrument is created by the English physicist William Crookes.

[26] Williman (1957, 39).

[27] Information discussed, as other Argentine authors claim that a few days earlier they would have conducted preliminary experiments at the University of Buenos Aires in which they obtained the radiographic image of a silverside. (Magazine, *Agriculture* , March 12, 1986, 430–433).

> Mr. Doctor Don Claudio Williman. My dear friend: I am interested in obtaining the radiographic photograph of the hand of this patient who says she has a needle stuck in the region. If it is possible for you to take care of the matter, either by taking that photograph or indicating who can do it, I will appreciate it. Your very affectionate Enrique Pouey, January 18, 1897[28]

The role played by the physics laboratory of Williman and Maggiolo was limited to experiments and some teaching for pre-university students, without the intention of using it as a medical application. But the Faculty of Medicine and some doctors felt the need to develop this type of studies in the Radiology Laboratory of the Charity Hospital (later renamed Maciel Hospital), existing since 1891. It was a photography and microphotography laboratory that managed to produce the first doctoral theses with radiographs in the country.

In the early years of the twentieth century, doctoral theses in medicine in Uruguay began to include the use of radiographs as part of the clinical and diagnostic documentation, a significant novelty for the time. According to Wilson and Wozniak,[29] among the first theses that integrated this resource stands out the one presented by Dr. Ernesto Morató in 1899, titled *Contribution to the study of coxa-vara*. This thesis, which obtained a "Regular" grade, includes a radiograph that documents the deformity of the femoral neck in a case of left unilateral coxa-vara.[30]

Another contemporary thesis is that of Dr. Horacio García Lagos, titled *Contribution to the study of acquired Genu Recurvatum*, defended in March 1900 and graded as "Good". This thesis also includes a radiograph taken in September 1899, used to plan a surgical procedure. The author details how he drew lines on the radiographic image to calculate the necessary angle for the intervention, reflecting an early use of radiographic images in surgical planning.[31] Finally, in December 1900, Dr. Juan Antonio Rodríguez presented his thesis *Coxa-vara consecutive to the false vicious ankylosis of the hip*, which also includes a radiograph taken at a local sanatorium. The first publications on radiology in the Medical Journal of Uruguay were also made.[32] Jacinto de León, mentioned for his links with physics from the field of medicine, will include in his courses the theme of this new phenomenon and its subsequent applications and improvements, from his chair of Medical Physics at the Faculty of Medicine.

The Institute of Radiology then emerged as a result of the initiative of the doctor and professor Américo Ricaldoni in 1912, who considered it feasible to manage the corresponding authorities public to establish in the Faculty of Medicine the Institute of Radiology. On September 6 of that year, President José Batlle y Ordóñez, along with Minister Juan Blengio Roca, expressed their support to the initiative of the

---

28 Williman (1957, 40).

29 Wilson and Wozniak (2020).

30 Ibidem, 30.

31 Ibidem, 31.

32 The Medical Journal of Uruguay in 1900 published a work presented to the Society of Medicine on July 27, 1900, titled: Encysted pleurisy on the left side whose author is Dr. Juan B. Morelli. This work features the first national radiograph published in the journal.

Faculty and presented to the consideration of the General Assembly a bill that allocated fifty thousand pesos for the acquisition of half a gram of radium and the equipment and installations necessary for its therapeutic application.

Carlos Butler, a 33-year-old Uruguayan doctor who trained in radiology in Germany, played a crucial role in the European mission in charge of acquiring radium and the necessary radiological material. Despite the limitations in the availability of radium, Butler managed to acquire 480 mg of this substance at the Usina Armet de Lisle,[33] France. Each unit of radium sold by the Usina was backed by a certificate of purity and quantity, signed by Marie Curie of the Paris Institute of Sciences. After the acquisition, the Institute of Radiology was established on December 16, 1913 at the Charity Hospital, thanks to a decree issued by the Executive Power:

> Article 1: The Institute of Radiology is hereby established in the Faculty of Medicine, in which the following will be practiced:
>
> a) Scientific studies on different radiations.
> b) Teaching courses on the same.
> c) Clinical applications of radiations to diagnosis and treatment of diseases.
>
> Article 2: The Institute of Radiology will have the following scientific staff, who will also be in charge of teaching: A Director. A Deputy Director. An assistant doctor.
>
> Article 3: The salary of the Institute's staff as well as the expenses that arise from its operation, renewal of materials, acquisition of new tools and devices, etc., will be covered with the income produced by the Institute itself.
>
> Article 4: The Institute will be installed in the Maciel Hospital and its operation and organization may be the subject of an agreement to be held between the Faculty of Medicine and the National Public Assistance.
>
> Batlle y Ordóñez, Baltasar Brum[34]

The creation of the Institute of Radiology initially pursued a dual purpose: on the one hand, it adopted a clinical approach oriented towards diagnosis and treatments, while on the other hand, it committed to a scientific approach, exploring radiological agents and providing education to medical students. Given the growing importance of the development of radiological science from a therapeutic perspective, the radiology section considered it essential to focus on this specialization at a national level. This implied not only practical application in the clinical field, but also the conducting of scientific research and the teaching of this medical branch.

The technical staff appointed by the Faculty of Medicine to work at the Institute included its director, Dr. Carlos Butler; the deputy director Dr. Mario C. Simeto; an

[33] The Armet de Lisle Power Plant is a French company established by Émile Armet de Lisle in 1904, specializing in the production of uranium salts and the manufacture of applicators derived from this element. Émile Armet de Lisle, a prominent French industrialist and chemist, played a crucial role in the development of the radioactive industry in France at the beginning of the twentieth century. In addition, he had significant associations with Pierre and Marie Curie and was the founder of the magazine Le Radium.

[34] Annals of the University (1912, 272).

assistant doctor, Dr. Eduardo Bastos, and a head of the Pathological Histology Laboratory, Dr. Federico Susviela Guarch.

With this milestone, Uruguay became the first Latin American country to incorporate this radioactive substance into its therapeutic practices. During these years, radiology was in an initial phase of experimentation and development, X-ray equipment was primitive, using simple tubes, and the quality of the images was limited. Radiology was mainly focused on basic diagnostic applications such as identifying bone fractures and locating foreign objects in the body. Health professionals were beginning to understand the risks associated with radiation exposure, and research was being conducted on the health effects. Radiology was becoming a distinct medical specialty, with radiologists exploring new ways to use the technology in the clinical field. Although the foundations for the use of X-rays in medicine had been established, the specialty was evolving, and advances in technology, imaging techniques, and radiological safety would occur in the coming years.

Teaching regarding the content of the training involved, according to the current curriculum, giving weekly or fortnightly classes to students in the different medical and surgical clinics, and giving an annual improvement course aimed at students and graduates. Doctors interested in this new specialty, in practice, learned with patients daily. To train in diagnosis they went to the Hospital; in therapy they attended the Institute of Radiology. Professor Butler and his assistants were in charge of teaching radiotherapy. The adjunct professors Cunha initially and Barcia later, took charge of teaching radiodiagnosis. In his performance, Butler defended the specific training of radiotherapy:

> We believe that the greatest successes are subject to the best and most complete preparation of each one, in such a way that one acts in an environment that brings together competent specialists in the various branches of medicine, because a radiology institute to be worth and produce, must gather or concentrate the greatest number of scientific competencies and the most efficient and complete material possible.
>
> Radiotherapy and oncology, subjects that given their current progress and their ever-increasing importance, should deserve the attention of the authorities and even be in charge of the obligation of their teaching, in a limited way for medical students and in a complete form for those who aspire to obtain the title of specialists.[35]

The program of subjects for obtaining the title of doctor, specialist in radiology, was developed through theoretical-practical teaching, 2 years long, consisted of four parts: radiological physics, radio-diagnosis, radio-biology and electro-radiotherapy and phototherapy. The curriculum of Butler's radiology course of 1915 covered various topics, among them: the nature of radiations, corpuscular radiations, cathode rays, roentgen rays, radioactivity, families of radioactive bodies, Becquerel and Curie rays, ultraviolet rays, infra and ultra red rays, cosmic rays, neutron radiation, radioactivity, the structure of the atom, roentgen ray devices, radiations and matter, filters, radiation measures for Roentgenterapia, devices for curietherapy, emanation,

[35] Historical Archive of the Faculty of Medicine (1944, 16–17).

measures for curietherapy, high frequency currents, diathermy, short waves, electrocoagulation, and the electric scalpel in association with radiotherapy.[36]

The sources of Butler's Radiology study programs do not provide a complete view of the depth with which the knowledge in physics present in them was addressed, nor are the bibliographic sources used detailed. However, considering Dr. Butler's trajectory as a professor, doctor, and researcher in the field, it is clear that there was a deep concern for the practical application of this knowledge. Although it is essential that radiology specialists understand the physical nature of radio, the importance of carrying out successful practices is even more emphasized. This medical knowledge is used in the diagnosis and treatment of various diseases, and daily radiotherapy lessons were carried out in close collaboration with patients during examination and treatment hours. Apparently, the most significant workload in training was dedicated to medical practices.

> The teaching activity was limited to two semester courses, which was considered insufficient by the director, and was especially aimed at teaching the therapeutic applications of radium. The care activity grew from 2913 patients in 1915 to 7163 in 1924.[37]

In several volumes of the Medical Journal of Uruguay (1915), Butler presents works such as "*Radium: techniques and indications for its treatment*,"[38] where he details real cases of *radium therapy* treatments and reveals the emphasis on the practical application of physical knowledge for the development of radiological research in the country. In addition, he addresses various topics in other writings and conferences, such as "Notes on the histological modifications produced by various radiations on normal and pathological tissues," "One hundred cases of cutaneous epithelioma treated by radium" and "On the histological examination of tumors as the basis of radiological treatment"[39] Butler's various publications reveal the contributions derived from his practical experience in the use of radium. It is important to note that Dr. Carlos Butler not only stood out as a doctor, but was also an enthusiastic radio amateur and active promoter of broadcasting in Uruguay.

After Butler's management, Dr. Pedro Barcia assumed as the new professor and director of the Institute of Radiology from 1945 to 1951. Focused on Radiodiagnosis and with a chair at the Maciel Hospital, Barcia introduced significant changes by renaming the institute as "Institute of Radiology and Physical Sciences". In addition, he established a Department of Medical Physics at the same Hospital, directed by Dr. Mario Cassinoni. During his 6 years of leadership until his death in 1951, Barcia organized prestigious annual radiological improvement courses, attracting the attention of foreign professionals. He also carried out a comprehensive restructuring of the Institute, incorporating a Medical Physics service, a discipline that was practiced in other specialties such as physiotherapy and physiodiagnosis. This medical branch focuses on the application of physical agents for healing,

[36] Butler (1915a, b, c).

[37] Wilson and Wozniak (2020, 50).

[38] Medical Journal of Uruguay (1915), Volume XVIII, 22–48.

[39] Medical Journal of Uruguay. (1915), Volume XVI, 442–463, 465–485 and 487.

encompassing not only radiations, as evidenced in the development of the Institute of Radiology, but also applications based on heat (thermotherapy), light (phototherapy), electricity (as developed in De Leon's classes), magnetism, atmospheric pressure (climatotherapy) and movement (Ling's Swedish Gymnastics).[40]

Physics in the activity of the Institute of Radiology was entirely subject to its practical utility in medicine. The history of the Institute of Radiology in Uruguay highlights the role of practical knowledge of physics in the advancement of medical radiology. Theoretical training in physics seems to be the minimum necessary to understand and handle devices for therapeutic purposes. There is almost no production on purely theoretical studies of phenomena such as radiation, which led-as we saw-those interested in physical phenomena, like Kayel, to question the relevance of such an Institute outside the Faculty of Engineering.

The work of the Institute of Radiology during these initial years focused on a pioneering phase of experimentation and development of radiology, with technological limitations, but laying the foundations for future advances. Undoubtedly, the teaching and application of radiotherapy, backed by research and daily practices, contributed to the growth of the specialty. The creation and evolution of the Institute of Radiology in Uruguay represents a prominent chapter in the convergence of physics and medicine, which we will see repeated in some subsequent decades. From its beginnings to the direction of Pedro Barcia, the institution consolidated itself as a center of innovation, research, and training at the intersection of these disciplines, laying the foundations for the continuous development of radiology in the country and the region.

3.7 Physicists and History and Philosophy of Science Circa 1930

Reaching the end of the second decade of the twentieth century, physics finds a handful of interested parties in the country, from two different perspectives: the medical application of radiology and the attempts at autonomous development of basic sciences in Engineering. This last and faint impulse will be crowned with the relevant role that the general diffusion of the new General Theory of Relativity acquires in the world.

The study of the reception and understanding of the general theory of relativity in Uruguay deserves to be approached from some necessary facts and questions.

As is also well known, in 1925 our country was visited by the most relevant scientist of the time (and probably of the entire twentieth century), Albert Einstein. This fact represented a significant revolution in the academic community of the University, if one sees it as one of the constitutive events for the development of theoretical physics in Uruguay. In fact, only if we travel to 1941, when the creation

[40] Iscardó (1923).

of the Institute of Physics of the Faculty of Engineering was approved, where (as its director, Walter S. Hill, communicated) it was intended to work in theoretical and nuclear physics, will we find traces of the cultivation of an autonomous physics. In other words, from 1916 to 1941, there do not seem to be traces of a moderately institutional development of physics within the Faculty. However, if we look at the activity of some of the young engineers who participated in academic life at the University, we will be able to see those sparks of light that can serve as a guide to understand how a modest corpus of knowledge in theoretical and experimental physics is being established (an element that we will study in the next chapter).

The question that may arise is the following: how is it that Amadeo Geille Castro can offer such a complete course, in 1923, on the theory of general relativity? How is it that by 1941 Walter S. Hill's production can already be known as someone who knew the rudiments of the new nuclear physics? How did these physicists emerge? How did they learn? Under what supports did they build their training? Did the influence of mathematics and the figure of Eduardo García de Zúñiga have anything to do with it?

It is likely to think that both Geille Castro, Enrique Loedel Palumbo or Walter S. Hill, were formed on that strong mathematical base that Eduardo García de Zúñiga so much promoted, and therefore it is convenient to start analyzing part of his legacy. As mentioned, to this original tradition of the old Faculty of Mathematics must be added the training of some doctors who attended both the theoretical and experimental advances of physics at the beginning of the twentieth century, especially in its medical applications. But to speak of tradition may seem excessive: within the Uruguayan academic life of the early twentieth century only a few were those who, in fact, understood the challenges that a professional scientific training meant for the new century. These few university professors sought to break with an encyclopedic formative tradition, based on a naturalistic philosophy that seemed totally outdated, in favor of a professional training like physics and international mathematics with high degrees of specialization. And if this small group of engineers took on the task of developing professional physics for the country, the name that deserves the credit for providing them with the conditions is that of García de Zúñiga. Before moving on to the new generation, we would like to highlight some personal aspect of the figure of Don Eduardo, but also of a figure little recognized in the history of Uruguayan intellectualism, who made his contributions to this new attention to science by this intellectual generation, this time from the field of history and philosophy of science, Paul F. Schurmann.

Eduardo García de Zúñiga was a well-known Uruguayan academic, both as the founder of a mathematical tradition in Uruguay and for his important role as an engineer in works and studies of relevance for the Batllista period. But in any case, in the memories that others have made of his professional and academic work, his value as a "humanist" is highlighted.[41] Without a doubt, García de Zúñiga was all that: an important actor in the national transformations of his time as Inspector

---

[41] Laguardia (1951), Cicalese (1992), Cheroni (1992).

General of Railways of the North of Río Negro, Director of Roads, Director of the National Ports Administration and Director of Railways. He was also an active participant in the transformations of the University itself: dean of the Faculty of Engineering, member of the councils of that Faculty and of the first governing body of the Faculty of Humanities and Sciences (FHyC), Professor *ad-honorem* of the Faculty of Engineering and doctor *honoris-causa* of the University. He presided over the Institute of Higher Studies, predecessor of the aforementioned FHyC. But what is most interesting is to suggest a different look: without disregarding his work as an intellectual of broad knowledge, it was his dedicated vision of what the professionalization of mathematics implied that has allowed the training of engineers who shared his vocation for mathematical and physical studies. Some facts help illustrate this circumstance little recognized in his career, because precisely the historiographic look has never stopped to perceive this aspect in detail. Eduardo García de Zúñiga maintained correspondence with prominent mathematicians of his time, to whom he sent works on matters that were the product of the innovation he introduced in his courses, as we have already seen. For example, from the result of the infinitesimal calculus course, he would have generated a text called *Note on the Wronskian*, on the introduction of a determinant for the study of ordinary differential equations. These works received the reading and comments of mathematicians such as Guido Fubini and Giuseppe Peano, when they were presented at a mathematics congress in Turin, Italy.[42] This type of link with foreign academic communities was very rare in the university environment of the early twentieth century, and should be seen as a substantive change in the ways in which academic legitimacy was built up to that point. To this aspect, the most relevant institutional contribution that García de Zúñiga made to the university must be highlighted, the formation of a library of regional recognition, composed not only of classic works of philosophy and Western thought, but with the subscription to international publications of mathematics. This way of conceiving mathematical teaching not only had an impact on the possibility of germinating a community of mathematicians who achieved important recognitions in the following decades, it also meant a substantive change in the way of thinking about science, not vaguely associated with the idea of culture, but as a fertile and necessary field for academic work. We must add to this impulse carried forward between 1910 and 1920, a second relevant fact: the confirmation of the general theory of relativity in 1919, the international fame of its author Albert Einstein, and the coincidence of his arrival in Montevideo in 1925.

In parallel with the initiatives promoted, García de Zúñiga was an active participant in the Uruguayan branch of the International Group in History of Science (a subsidiary of the International Committee for the History of Sciences, in Paris). His connection with the mathematician Julio Rey Pastor, responsible for the creation of this group, along with the Italian Aldo Mieli, made it possible for this group to find an expression in Uruguay, within the Institute of Higher Studies (which depended on Udelar). Within this group, it is necessary to highlight the Belgian professor Paul

[42] Laguardia (1951).

Schurmann, of Belgian origin, who served as a professor at this Institute. The need to do so has to do with a monumental work on the History of Physics, the first major work published in Latin America, which was prefaced by García de Zúñiga himself.

> The "History of Physics" of my friend Professor Schurmann has, indeed, in turn, a small history that will partly explain the spontaneity of my role as a preface writer. Started many years ago and presented in 1925 to an open competition by the Ministry of Public Instruction, it was proposed for the prize of the best scientific work by the corresponding tribunal for mathematical and physical sciences. The chance of being part of this tribunal and the easy defense that I took before it of Schurmann's work, until its publication by the University was decided, are precedents that already confer on me a sort of godfatherhood, too honorable to renounce now that I can make it public.
>
> And there is another reason, even more personal, that spiritually attracts me to Schurmann's beautiful work; and it is that I myself have been cultivating for some time, and with great enthusiasm, the history of mathematics, so closely linked with the history of physics that it confuses with it in more than one chapter.[43]

The work, with more than 1300 pages, addresses in great detail the history of universal physics from the reconstruction of its most relevant heroes, making them a narrative between biographical and scientific. Published in 1936, the book stops precisely at the point where modern physics begins the most important contributions and discoveries, from the twentieth century. This is one of the aspects that García de Zúñiga highlights as one of the two most fallible aspects of the work: not delving into the most recent problems that physics presented in those years; as well as not having a more updated bibliography of these more current issues of modern physics, which would have been of interest and necessity for young students. But without a doubt this work marks a milestone of relevance not only for Uruguay but for the Latin American context, as evidenced by the critical reception international that it received.

The French philosopher Hélène Metzger Brühl defined in these terms Schurmann's work:

> ... Here is in Spanish a history of physics that the general public will read with profit and joy, as much as the specialists who will find in it an infinity of information on different issues than those they are laboriously trying to elucidate. This great work of more than 1300 pages will truly provide outstanding services because it will constitute a background of the activity of human thought in its eagerness to represent and dominate the world around us.
>
> Mr. Schurmann has made a chronological study of the work of the greatest men of whom humanity has been able to take pride and his exposition includes the scientific biography of the illustrious physicists from antiquity to our days. He quickly studies the wise men, the oldest and those of the Middle Ages, he dwells more at length on those of modern times and finally gives a predominant importance to the contemporary period that goes from the beginning of the nineteenth century to the present time.
>
> The history of physics is therefore in perspective as befits a contemporary thinker who, granting today's science, which he knows perfectly well, the first place, does not forget that

---

[43] Schurmann (1936, III).

> this science results from the more or less active collaboration of men of all times and all countries.[44]

Among several reviews that are gathered in the book that Schurmann later publishes on the origin and concepts of thermodynamics, appear reviews from physicists like Jean Perrin and Louis De Broglie, where the latter writes:

> ... This vast work so extensive and so erudite is of the highest interest and I continue its reading with maximum pleasure and the greatest benefit for my personal instruction. It is a beautiful work that took a lot of time and that will surely be very useful for historians of science and for physicists.[45]

At a time when the construction of physics in Uruguay was beginning, as has been attempted to demonstrate in this chapter, the history of the discipline carried out by a Belgian professor in the Uruguayan territory also meant a contribution of support towards the need for an autonomous development of the discipline. Schurmann, although trained as a physicist but turned historian, became an ally of the emerging and small scientific community that was born in the country, under the leadership of his friend, García de Zúñiga. In 1941, Schurmann gave a lecture called *The problem of culture*. This text is of great interest as it indicates some relevant data for Uruguayan science. One of them is that it affirms that García de Zúñiga would have met Einstein in the United States, a fact that has not been able to be confirmed with the existing documentation of the physicist:

> The United States has taken advantage of the terrible European crisis and others previous ones to receive, generously yes, but also with a healthy and foresighted selfishness, all the wise men, thinkers and artists rejected from the old world by the horrendous windstorm. And there is Einstein to not choose but one, but wonderful example, Einstein, whom Don Eduardo Garcías de Zúñiga visited in the United States and who did not want to speak to him but in English, as a sign of gratitude, love and admiration for the great American people.[46]

The second aspect of this conference that deserves attention has to do with glimpsing the problem of specialized, professional culture, within the bosom of university institutions. Otherwise, how will universities cultivate a generalist education, promote "genius" from the material conditions required by the new science. In this sense, Schurmann's feeling is one of skepticism like the one that augurs, by those dates, the Spanish philosopher Ortega y Gasset, of undoubted influence on Uruguayan intellectuals:

> Ortega y Gasset has written the following lines: "The University contemporary has greatly complicated professional teaching; has added research, completely removing the teaching or transmission of culture. This has been an atrocity. Dire consequences of this that Europe now pays, because the average Englishman, the average Frenchman, the average German, are uncultured, they do not possess the vital system of ideas about the world and man corresponding to the time.

[44] Schurmann (1938a, b, 92).

[45] Schurmann (1938a, b, 94).

[46] Schurmann (1936, 94).

> This character is The New Barbarian, delayed with respect to his time, archaic and primitive.
>
> The new barbarian is mainly the professional, wiser than ever, but also more uncultured: the engineer, the lawyer, the doctor, the scientist."[47]

The professionalization of science, according to the diagnosis of not a few intellectuals, far from expanding the horizons of knowledge, leads them to an increasingly narrow limitation. It is an international tension that will see its manifestation in Uruguay in these first decades of the twentieth century, accentuated by the irruption of relativity and the visit of Einstein. In 1987, the physicist and philosopher of science John Ziman, caricatured this tension between science and philosophy that underlies the type of tensions mentioned by Schurmann:

> A philosopher is a person who knows less and less about more and more, until he knows nothing about everything
>
> A scientist is a person who knows more and more about less and less, until he knows everything about nothing.[48]

Bibliography

Annals of the University. 1912. Year XVII, Volume XXI, 39–40.

Butler, C. 1915a. *Medical Journal of Uruguay (1915), Volume XVI, XVIII*, Montevideo: Imp. The illustrated century of Gregorio. V. Mariño.

Butler, C. 1915b. *Communication to the Society of Medicine of Montevideo, Session April 9.* Udelar: Historical archive of the Faculty of Medicine.

Butler, C. 1915c. *Radiology Course*. Udelar: Historical archive of the Faculty of Medicine.

Cheroni, A. 1992. Eduardo García de Zúñiga. A humanist militant. Montevideo: Division of Publications and Editions. *University Gazette* 6 (1): 37–38.

Cicalese, V. 1992. Eduardo García de Zúñiga, a consummate humanist. *University Gazette* 6 (1): 44–45.

Council Minutes. 1914. *Council Minutes Faculty of Engineering., Book III*. Archive of the Faculty of Engineering, University of the Republic.

Council Minutes. 1916. *Minutes of the Faculty of Engineering and Related Branches Council, Book III*. Archive of the Faculty of Engineering, University of the Republic.

Finocchiaro, A. 2013. *The Reformist Myth*. Buenos Aires: EUDEBA.

García de Zúñiga, E. 1907. *Journal of the Association of Engineers and Architects* 6:August–September.

García de Zúñiga, E. 1914. *Study on Preparatory and Superior Mathematics Programs*. Montevideo: Library of the Journal of the Polytechnic Association of Uruguay.

Historical Archive of the Faculty of Medicine. 1944. *Summary of the work of Professor Dr. Carlos Butler*. Historical Archive of the Faculty of Medicine, Montevideo, Florensa y Lafon Edition.

Iscardó, V. 1923. *Program for the Physiotherapy Course*. Montevideo: Historical Archive of the Faculty of Medicine.

Laguardia, R. 1951. Eduardo García de Zúñiga. Some aspects of his personality. *Journal of the Faculty of Humanities and Sciences* 7.

---

[47] Schurmann (1936, 95).

[48] Ziman (1987, Forewords).

Medical Journal of Uruguay. 1913. *Medical Journal of Uruguay, Vol. XVI & XVIII*. Montevideo: Impresiones El Siglo Ilustrado.
Monteverde, J. 1907. *Journal of the Association of Engineers and Architects* 1:85.
Monteverde, J. 1915. *The Teaching of Mathematics in Engineering*. Montevideo: La Razón, El Siglo and El Telegraph Workshops.
Oddone, J., and B. Paris de Oddone. 2010. *History of the University of the Republic*. In *The University from Militarism to the Crisis. 1885–1958*, vol. 2. Montevideo: Editions of the University of the Republic, University Classics Collection.
Piaggio, N. 1877. Parallel between Philosophy and Mathematics. In *Bulletin of the Society of Sciences and Arts, Montevideo*, 189–190.
Quijano, C. 1941. The university reform and the Faculty of Law of Montevideo. In *The University Reform. Volume III. Critical Essays (1918–1940)*, ed. G. Del Mazo, 223–233. La Plata: Edition of the Center of Engineering Students.
Renaissance. 1926. *Monthly Student Newspaper* 7:72.
Schurmann, P. 1936. *History of Physics*. Vol. 136. Montevideo: Annals of the University.
Schurmann, P. 1938a. *The Birth of Thermodynamics*. Uruguayan Group of History of Science, Imp. Helios.
Schurmann, P. 1938b. *The Problem of Culture*. Montevideo: Modern Imp.
Stillwell, J., and J. Stillwell. 1989. *Mathematics and Its History*. Vol. 3. New York: Springer.
Thomasset Trakalo, J. P. 2019. Bernardo Kayel, Wireless Telegraphy Engineer. *Engineering Journal* 84:10–15.
Waiter, A. 2019. *Technological Trajectory, National Capabilities and Institutional Aspects: The Construction of the Hydroelectric Dam in Rincón del Bonete, Uruguay*, 1904–1945.
Williman, J. C. 1957. *Dr. Claudio Williman, His Public Life*. Montevideo.
Wilson, E., and A. Wozniak. 2020. *Diagnostic Radiology in Uruguay. A Historical Approach*. Montevideo: Society of Radiology and Imaging in Uruguay.
Yaffé, J. 2000. Ideas, program and economic policy of Batllismo. Uruguay 1911–1930. In *Working Paper Series; 07/00*.
Ziman, J. 1987. *Knowing Everything About Nothing: Specialization and Change in Research Careers*. Cambridge: Cambridge University Press.

Chapter 4
The Theory of Relativity and Its Cultural Impact on the University

The theory of relativity has produced countless works of various magnitude, scope, and themes. The history of science and the philosophy of science of the twentieth century cannot ignore the impact of the Einsteinian theory, whether to praise its revolutionary character or to highlight its inconsistencies and problematic nuances. This chapter aims to show the context that awakens the reception of the theory in the different spheres of scientific and philosophical cultures. This type of studies, which have spread throughout the world, marks the different impacts that the theory had on various academic communities, mainly in the communities of engineers, mathematicians, biologists, and, of course, philosophers.[1] In the Uruguayan case, it is intended to account, above all, for the immense complexity that it meant to assume, understand and work on the new Einsteinian physics, which posed a challenge in several dimensions. First, we can highlight the intrinsic complexity of the general theory, a work that sought to generalize the results obtained by Einstein in 1905, when he published his series of articles among which the theory of special relativity appears.[2] This means something very simple, those who approach general relativity from its confirmation by the experiments carried out by Edington in 1919, generally ignore the entire history of development, finding themselves from one day to the next before a new possibility of explaining the principles of physics that is extremely abstruse. Secondly, the new Einsteinian theory required a series of knowledge that involved disciplines that, except in a few advanced spaces, remained separate. The construction of this general framework in which the theory resides, a framework built by Einstein's genius, was difficult to incorporate within the academic traditions of the time. And this difficulty was increased in the situation of countries like Uruguay, where-as we saw-scientific development was almost nil. Thirdly, there have been everywhere a series of philosophical readings on what the new relativity

[1] Krauss and Laino (2023).

[2] Holton (1973, 1978).

J. A. Queijo Olano, E. Mozo Meneses, *Giants Upon Our Shoulders: A Philosophical History of Physics in Uruguay*, Studies in History and Philosophy of Science 65, https://doi.org/10.1007/978-3-031-97729-9_4

proposed, which established dialogues and proposed readings that managed to settle in academic communities, despite not always being entirely correct or fully accurate in their interpretation. From the empiricism *of Mach* to a vitalist existentialism like that of Bergson, philosophy was a recognized interpreter in various contexts, especially in those where the scientific specialization required to understand the new great scientific milestone was not yet fully developed. This last form of reception, with a central role in philosophy, seems to have played an important role in Uruguay if we consider the role played by Carlos Vaz Ferreira during the visit of Einstein to Montevideo, one of the most prestigious intellectuals of the country.

4.1 The Philosopher Carlos Vaz Ferreira, Between Einstein and Bergson

In 1922 Vaz Ferreira would have given a talk on the theory of relativity at the Ateneo de Montevideo.[3] The physicist Loedel Palumbo, who was studying engineering at that time, remembered that talk in a 1955 article published by the Engineering Journal, of the Association of Engineers of Uruguay, in tribute to the death of Einstein:

> Back in 1923 [the conference would have been in 1922], I had the pleasure of listening to a conference by the Uruguayan thinker Carlos Vaz Ferreira on relativity. With the intellectual honesty that characterizes him, he spoke of the barbed wire fences that constituted for him the symbols of tensor calculus and that prevented him from entering the fortified redoubt of the theory. And he, without being a physicist, understanding the deep cultural significance of it, tried to unravel from that sea of incomprehensible formulas, the ideas that, necessarily, had to constitute the basis of the imposing building. And the ideas are simple, extremely simple, and are within the reach of any common mentality.[4]

Many years later, the philosopher Otero[5] marks the same idea about the mathematical difficulties of Vaz to be able to understand and discuss the Einsteinian theory, difficulties that are recognized in a certain sense by Vaz himself in his text *Illegitimate*

[3] The Ateneo de Montevideo is a society aimed at promoting the cultural and intellectual development of the Uruguayan capital, which has its antecedents in the Ateneo del Uruguay and the University Society, previous groups with similar purposes. Mario H. Otero notes that existing documentation suggests that this talk by Vaz Ferreira would have been given in this space (Moraes 2019). In any case, this and subsequent approaches to physics topics, such as quantum mechanics, led to the development of Vaz's work *Illegitimate Mathematical Transcendentalizations*, from 1940. Otero's criticism of Vaz stems from the claims made in this text.

[4] Palumbo (1955, 73–74).

[5] Mario H. Otero (1929–2013) was a Uruguayan philosopher, founder of the Department of Philosophy of Science at the FHyC, in 1964. A graduate in Philosophy from that Faculty, he continued his studies at foreign universities such as Berkeley and Buffalo (United States). His doctoral thesis, *The contribution of Boole and De Morgan to formal logic*, he obtained at the University of Paris. Through a Fulbright scholarship, he studied at Harvard with philosophers such as Willard Van Orman Quine and Hilary Putnam.

mathematical transcendentalizations and correlated fallacies (Vaz Ferreira 1940). This book had been constituted from some conferences that Vaz Ferreira gave in 1939 at the UBA, where he would have inquired about the idea of the mathematical transcendentalization in some cases of the most contemporary sciences, these are, the theory of relativity, non Aristotelian logic and Heisenberg's uncertainty relation (known as the uncertainty principle). As Otero points out, these works would have been recovered from some conferences given on these topics, but they still show clear signs of the relative "delay" with which the country's leading philosophical exponent approached these topics:

> The theory of relativity is from 1905 (the restricted one) and from 1907–1913 (the general). The Heisenberg principle of 1926. Non-Aristotelian logic is from 1877–1878 in Mac Coll, from 1911 to 1924 in Vasiliiev, from 1920 in Lukasiewicz. This does not take away from the scientific-philosophical relevance of these topics, which are still current today, but it gives an idea of the delay with which they are disseminated in the Uruguayan philosophical world. Vaz Ferreira takes up relativity at the beginning of the twenties, quantum mechanics (1934) and non-Aristotelian logic (1938) around the mid-thirties.[6]

Vaz Ferreira's work from 1940, presumably the result of that intervention in 1923, deals with the theory of relativity. We believe it would not be entirely fair to judge Vaz Ferreira's positions as "backward" in the face of a complex theory that was quite recent enough to be understood by an audience unfamiliar with physics and mathematics. Perhaps it is better to understand some of his reasoning in light of an even greater conflict, such as the one that arose between Einstein and Bergson in that same 1922. In any case, it is advisable to start by showing how Vaz Ferreira interpreted some key concepts of the Einsteinian theory:

> When the theory of relativity arrived in these countries, I gave a long series of lectures to address these aspects, which, naturally, I cannot—nor would it be timely—summarize here; but what has to do with mathematical transcendentalizations was essentially the following (very synthetic):
>
> Relativity defines—initially— simultaneity as perceived simultaneity (simultaneous, what is perceived as such, as occurring at the same time); and referring to perception by light.
>
> On the other hand, classical theories, starting from an absolute space in itself, and an absolute time also in itself—postulating it thus—defined simultaneity as in itself, disregarding the perception of it, and, consequently, also, of possible necessary agents and the time that might be necessary for that perception (this has nothing to do with what was known or not known about it at the time).
>
> So, there are two definitions of simultaneity. One: what would be simultaneous without the intervention of agents or means. Another: what is perceived as simultaneous.
>
> Well: both definitions are indifferently legitimate—as such—(understanding about the meaning, and keeping consequence). But if they are transcendentalized, then there would be a contradiction (which no one would fail to perceive in the same definition, but not so easily in the consequences and developments, where it would appear disguised). If they are transcendentalized, there would be a contradiction (that is, they would appear in contradiction, since, precisely, it is illegitimate to transcendentalize definitions).
>
> However, it was a psychological process that was difficult to avoid... The philosophy of classical mechanics contained a transcendentalization of the concept of simultaneity. While

[6] Otero (1996, 52).

> the theory of relativity brought (I tried to explain it then) a transcendentalization of the perception of simultaneity (and of a special perception: visual perception).[7]

There is no doubt about the confusion that Vaz enters when trying to interpret the idea of simultaneity in relativity, and which he tries to summarize in the Latinism Esse est percipi per lucem ("to be is to be perceived"). Firstly, simultaneity in the Einsteinian theory is a problem that must be solved when entering the certainty that there are neither absolute space nor absolute time, and that these dimensions can only be measured within established reference systems, which are always relative to others. But this, as Otero points out, is not a problem that should be thought of in terms of perception (less of visual perception).

The other great aspect that shows a dissonance in the reception of relativity elaborated by Vaz Ferreira is the ontological character he gives to perception, when he establishes:

> But if these new definitions, resulting concepts and related mathematical formulas are transcendentalized, by pretending to make them realities, then we are going to believe, through this process, that there is, between the classical doctrine and the relativistic doctrine, a greater contradiction than the one that actually exists.[8]

The other big problem that Vaz Ferreira points out is what he calls transcendentalizing, that is, granting a reality character to the concepts and definitions established in the general law of relativity. Here lies something deeper, which has to do with the explanatory power of laws and the ontological character of some concepts. Vaz Ferreira seems to want to tell us that the new physics may offer better explanations and open new possibilities for the future, but that the big problem we can have is to transcendentalize its basic postulates, that is, to believe that reality is as described by the law. If we do not perform this operation, what we can understand is that Newtonian physics is only surpassed by Einsteinian physics because it has less predictive power and possibility to explain new phenomena, not necessarily because it describes physical phenomena better.

Without a doubt, Vaz Ferreira's criticism of this notion, as well as He will later do so with Heisenberg's principle, inherently carrying the idea of a science that can give us better explanations about phenomena but not establish the ontological reality of what surrounds us. But these ideas are far from being exclusive to the Uruguayan philosopher. They are framed in a great debate that had arisen in that 1922 between Einstein and Bergson, and which Vaz Ferreira followed closely.

Without delving into the details of the episode or its antecedents or consequences; some relevant aspects must be attended to understand what happened with Vaz Ferreira and his lecture on the theory of relativity of 1922.[9] That same year, on April 6, Einstein is invited to the Collège de France to give a lecture on his theory, organized by the Société Française de Philosophie. Henri Bergson, who by that time was

[7] Vaz Ferreira (1940, 12–13).

[8] Vaz Ferreira (1940, 12–13).

[9] For an extensive and dedicated understanding of the tension between Einstein and Bergson, we recommend the work of Jimena Canales, *The Physicist and the Philosopher* (Canales 2020).

a highly distinguished philosopher in all parts of the world, would have an uncordial encounter with the new intellectual model of the twentieth century, the theoretical physicist. The dispute, in which we will not delve into details, will be established from the concept of *time*. On the one hand, Einstein defended the objective position of his relativistic theory of space-time and the function that simultaneity fulfilled in it; on the other hand, Bergson defends this legendary concept in terms of an "élan vital"—a time as vital duration—saying that the theory of relativity ventures into philosophical territories and moves away from the purposes of physical sciences. And in a lapidary way, what resonated most from that encounter has been Einstein's sentence that stated: "the time of the philosophers does not exist."[10]

Vaz Ferreira seems to have been aware of the dispute, without being entirely clear about his final position on one or the other position. As an example, in 1925-the year of Einstein's visit to Latin America-he gives a lecture on Bergson of which some notes are preserved.[11] There he recognizes some aspects of Bergson's philosophical form with which he shows distance, using an example that has many implications in relation to his differences with Einstein

> Perhaps more reproach could be made for something else: For a certain tendency to follow certain systematizing ideas too much, if you want for too much obsession with these systematizing ideas. It occurs to me as an example something that we have already analyzed: That refutation of Zenon of Elea's argument about movement, based on the fact that Zenon, in his alleged demonstration, would have disregarded the fact of life. The steps of Achilles and the steps of the turtle, Bergson tells us, are living things, and therefore, they are indivisible. Dividing them would be, according to him, the result of the illusion (which for Bergson constitutes the great illusion of modern philosophy and science) of representing life mathematically...
>
> and this is so fictitious, that one would not know how to apply them to the case of that, in movements, there was no living thing, nor, therefore, indivisible and in the case, for example, of two bullets fired with different speed, the one behind could not reach the one in front and Zenon could not be refuted...[12]

It is undeniable that Vaz Ferreira does not feel comfortable with the proposal to reduce all movement to a "vital form" thus disregarding the advances that mathematization built in physics since the modern era. Bergson constructs the struggle on the idea that these processes of "objectification" that modernity built on logic and mathematics, are nothing more than mere constructs, where the theory of relativity would be its last expression. But Vaz Ferreira seems to want to concede something to the physical world that is beyond this vitalist conception, although without expressing it explicitly.

> Time is for Bergson the supreme reality, while space is a kind of scientific fiction. The only thing that is expressly discussed in Bergsonian philosophy, is whether time is as he presents it, real and indivisible. But will space be so different in nature, so fictitious, so purely scientific? The recent controversy between Bergsonians and Einsteinians has shown that there

---

[10] Canales (2020, 35).

[11] Vaz Ferreira (1972).

[12] Vaz Ferreira (1972, 76).

> was here, where I announced it, a vulnerable point. A certain theory derived mainly from Minkowsky, would tend to give time and space to such a point the same character that they would perhaps form a single reality.[13]

But what did Vaz Ferreira really consider about relativity? Why was he the one in charge of representing the country's intellectuality when Einstein arrives in Montevideo? Some anecdotal considerations have surrounded all the myth built on Einstein's visit to Uruguay, and which have been sufficiently collected in articles and popular science books.[14] For our purposes it seems relevant to consider two things: the existing material related to Vaz's approaches to the theory of relativity, and also, the reception in another set of intellectuals in the country who published about it.

While the figure of Vaz Ferreira has been, in popular imagination, the one most associated with Einstein's visit to Uruguay, a deeper history of the actors concerned with physics in the country should remove the excessive focus on this intellectual. For example, the introductory study to the edition of *The Travel Diaries of Albert Einstein. South America, 1925*, by Ze'ev Rosenkranz, the only thing said about the context of Uruguay in the year of the German physicist's visit, is the following:

> In contrast to the situation in Argentina, Uruguay and Brazil did not have such infrastructure in the exact sciences and therefore the reception of relativity was delayed there. At the time of Einstein's visit to Uruguay, physics was only studied at the University of the Republic's School of Engineering and at the Polytechnic Association of Uruguay.[15]

Without wanting to exaggerate what were undoubtedly isolated, sporadic and very personal impulses, there were some university actors who paid attention to this new great theory. Even more, some of them-we can infer-, allowed to think about possible developments of theoretical physics where there were none, precisely from their interest in relativity.

4.2 Amadeo Geille Castro and the First Paper on the Theory of Relativity

Amadeo Geille Castro played a central role in the reception of relativity in Uruguay, in a double literal sense: he was the one who taught a complete course on the theory and published his lectures in 1923, officially providing the University with its first material on relativity. In a second sense: he was responsible for going to fetch Einstein in Buenos Aires and accompanying him to Montevideo, on behalf of the Faculty of Engineering. Although the official reception of Einstein has been very

[13] Vaz Ferreira (1972, 79).

[14] Cabeza et al. (n.d.); Moraes (2019).

[15] Rosenkranz and Einstein (2023, 6).

well documented,[16] the question remains about who was this engineer capable of giving such a course, when there was no reference or previous work on the subject at the University.

At the time of the course, Geille Castro held the chair of Rational Mechanics, one of the subjects most marked by knowledge in physics in the engineering career. Now, Geille Castro's training has a lot to do with his family context, where all his brothers played an important social role in Montevideo at the beginning of the twentieth century.[17] In his case, in addition to completing his career as an engineer, and being a teacher of Rational Mechanics, his vocation for mathematics, which led him to hold the chair of algebra in the Secondary Education Section of the University, to give courses on *extension of mathematics* at the Uruguayan College-an institution authorized by the University to reinforce the "elementary, university and commercial" education, and to write some important works when he was an advanced student. While still a student, Geille Castro regularly collaborates in the publication *Evolution*, a student magazine that reserved not a few pages to works elaborated by students still in career. In this medium, he publishes in 1907 a *Study on the precision of the equinoxes*,[18] an astronomical study on the measurement of the phenomena of movement of the equatorial line with respect to the position of Aries, which is the historically used reference for these studies. Geille Castro shows not only how the effect is produced, his calculations with the available elements, but some of the consequences of these celestial movements and their impact on different dimensions of science. In that magazine, a few years later-in 1911, he will serve as secretary of the section *Scientific Investigations*, of which the French engineer Félix Pernot.[19] The data is relevant, because this section establishes a fairly strict regulation for the publication of academic works, especially those in exact and applied

[16] In this regard, the book *Einstein in Uruguay*, by Moraes (2019) is recommended. It provides an account of the cultural state that existed in Montevideo in those years, with a particular interest in the reception that the press made of the visit.

[17] The Geille family, originally from the Saint-Léger area of France, only knew South America when Amadeo Geille, the engineer's father, arrived in Brazil having been orphaned as a child. At that time, he worked as an agent for the company "J. Albert e Comp", which exported coffee from Rio de Janeiro to Le Havre. In the 1860s he married Laura de Castro, daughter of Don Agustín de Castro, a legislator of the Republic. The couple had several children, all with a prominent political, cultural and social activity: Enrique, a law doctor and official of the Supreme Court of Justice; Amadeo, the engineer; Agustín, who was consul of Uruguay in Turin from 1919 to 1926; Luis, owner of a jewelry store; María, who would marry the doctor Pedro Visca Peluffo; and Renée Geille, a prominent plastic artist who married the jurist Rodolfo Sayagués Laso. Via https://geilledesaintleger.wordpress.com/les-geille-damerique-du-sud/, accessed on 15 February 2024.

[18] Geille Castro (1907).

[19] Émile Martin Félix Pernot was a retired captain of the French artillery, who graduated from the Polytechnic School of that country, where he also served as a professor of higher mathematics in Paris schools—there he wrote a booklet for high school students on descriptive geometry. Upon his arrival in Montevideo, he became director of mathematical studies at the Military and Naval School.

sciences.[20] It is about erecting a section with certain academic rigor, destined to scientific research, especially in mathematics. This reinforces the indications suggested above: the changes in focus that occur in the teaching of mathematics at the University, promoted mainly by Eduardo García de Zúñiga, begin to show their influence on some students who, a few years later, will be the first engineers to develop a thought oriented to theoretical physics.

In 1910, Geille Castro publishes in *Evolution* a work called *Interesting application of an ancient theorem*,[21] on Ptolemy's Theorem, for those students of extension of mathematics. In 1911, he publishes a work on *Newton's binomial*, a work of great relevance that deserves the approval of the Annals of the University to be published:

> This new production of Mr. Geille studies the so interesting binomial, presenting it to us in such a complete didactic form, that the University found it useful to adopt it as a text and publish it in The Annals.
>
> All mathematics students at our University, will find in Newton's "Binomial", the most complete development of that exam topic.[22]

All this activity and work in the field of mathematics show us a young student who, by the 1920s, has already become an engineer and has managed to establish himself

---

[20] Here we will highlight some of the most relevant articles of this *Regulation*:

Article 1.° The EVOLUTION magazine creates the "*Scientific Research Section*", whose purpose is to stimulate intellectuals, especially students, to create, innovate and modify scientific topics.

Art. 2.° The section will receive contributions on any of the so-called "exact" sciences (pure and applied mathematics), and will publish them, accompanied by the name of their author, as long as there is something new and of merit in them.

Art. 3.° It is not required to publish a work, that it constitutes an invention; all those fundamental modifications on things already demonstrated, all those applications of old principles, and all those new demonstrations on known theorems, that reveal ingenuity. These works may deal with elementary or superior topics.

Art. 4.° The section will also propose Mathematics problems, ordered according to their difficulty, and the best solution for each one will be published monthly, with the name of its author appearing. In the case of believing it convenient the Direction, more than one solution of a same problem will appear. The names of the authors of good solutions that are not published will also be announced.

Art. 5.° To judge solutions, not only the demonstration of mathematical knowledge will be taken into account, but also the elegance of the exposition, the ingenious, practical and simple of reasoning.

Art. 6.° For the election of the best solution, the capacity of the sender will also be taken very into account, always giving preference, under equal conditions, to the solution sent by the collaborator who, apparently, may possess less knowledge. (…)

Art. 12.° Foreign collaborations will be admitted, with articles 10 and 11 applying to these. (…)

Art. 15.° Works from the students of the Military and Naval School, on sciences that are taught in that Institution, will be admitted.

Art. 16.° The Direction will especially appreciate the sending of problems of a scientific nature, that arise in the exercise of the careers of Engineer, Surveyor, etc., believing that this constitutes a procedure to ensure that these practical application problems are studied in a complete manner (Pernot and Geille Castro 1911).

[21] Geille Castro (1910, 138).

[22] Evolution (1911, 75).

as one of the few scholars in physics topics, which led him to offer the first course and be author of the first complete work on the new theory of relativity, and he is the university representative to go to fetch Einstein in Buenos Aires, share his academic activities there, and escort him to Montevideo in 1925. Let's analyze in detail both events.

In the publication of his course on Theory of Relativity, the dean of the Faculty of Engineering, Eng. Donato Gaminara, defines Geille Castro in the following way:

> (...) Professor Geille is one of our colleagues who has most dedicated himself to mathematics and Mechanics, so his word is one of the most authoritative to address the topic.[23]

His work will be the most significant that will be developed in the country on the subject, prior to Einstein's arrival in the country. The work of Geille Castro has 140 pages, when in that period the exercises to interpret and discuss the theory barely exceeded the thirty pages. Geille Castro begins his work trying to explain the conditions of the reception of the theory of relativity in Uruguay, indicating that he finds two fundamental reasons: ours is a "young" nation, "more attentive to the conquests of industry than to the discoveries of science", and on the other hand, being located far from the European centers where these discoveries are produced, "the news and books reach us with a considerable delay."[24] The other aspect that the author highlights, has to do with the public understanding of the theory itself, which is seen in two central attitudes, ranging from those who assume that the theory can only be understood by a handful of people, to those who claim to understand it just by reading pamphlets and very short press articles. To this panorama, Geille Castro responds by quoting a phrase from the German mathematician Felix Klein: "The conceptions of which we are concerned are the product of a prolonged work of mathematical thought and are far removed from the thoughts that are commonly used in life."[25] Indeed, the complexity of the theory lies in the fact that it is not only a complex construction that has its history, but also that history is not due to the knowledge of one discipline, but of many, "it is a theory of such enormous, so universal projections, that it encompasses almost all of the so-called exact sciences. Thus, for its understanding a quantity of knowledge from the most diverse sciences is needed, difficult to gather in one person, without a preconceived end."[26]

The end of the first of his lessons on the subject rescues an essential aspect of the role that mathematics played in the reception of relativity in Uruguay. There Geille Castro warns that what will be seen in the following meetings is not a historical reconstruction of the theory, but its mathematical manifestation, the only one that enables its real understanding. In a way, what Geille tries to convey is not only his mathematizing vision of physics (what Vaz Ferreira sees as an "illegitimate transcendentalization"), but also there lies the only corpus of knowledge more prepared

[23] Geille Castro (1924a, b, 75).

[24] Geille Castro (1924a, 7–8).

[25] Geille Castro (1924a, 8).

[26] Geille Castro (1924a, 8).

in the university to analyze a discovery like that of the theory of relativity. Physics can only be understood as theoretical physics if mathematics have a central role.

> As can be inferred from this dissertation, the theory of relativity has had an essentially physical principle. Physicists have been the wise men who have intervened in its formation, physicists the experiments that have produced it. However, this theory applies afterwards to all exact sciences. To avoid a series of previous knowledge and to be more in line with the nature of the studies that are pursued here, we are going to present it, not as it was engendered, but in a mathematical form: we will base ourselves only on mathematics and mechanics to deduce the theory. It was necessary that those who are going to attend these lessons knew that the doctrine was not formed historically like this, but that once engendered by the indirect path of physics, it has been able to be explained in a mathematical and mechanical form, leaving aside its origin, to make it easier.[27]

Even so, the set of the eight lessons that make up this first Uruguayan "treatise" on the theory of relativity, not only shows a capacity for study and understanding of the complex of aspects that make up the theory, but a deep understanding of how equations and experiments work, how the dialectic between researchers in the field of optics (Bradley and Michelson-Morley) has taken place to to determine the nature of the ether; how non-Euclidean geometries were constructed—up to Einstein's use of Riemann's work—from the historical analysis to Euclid's fifth postulate on parallels. Geille Castro handles a set of metaphors and examples that more than adequately illustrate each of the points that make up the theory of relativity, without thereby failing to reach the arid mathematical manifestations of these problems.

And not only did he provide a complete course on the Theory of Relativity (the restricted and the general, their differences and their respective scopes), but in the last lessons he showed how the theory was proven, what experiments worked for it and what type of measurements provided the evidence that the community of physicists needed to legitimize the discovery. And even more, he found space to show the problems that remained open (for example, to unite laws of electromagnetism with universal gravitation) and the momentum that the mathematician Hermann Weyl was pushing forward.

To conclude, it is worth rescuing some final notes on the analysis that Geille Castro makes of the theory itself, always in a modest and cautious way. We find it important to focus on two central aspects. One has to do with this discussion that appears in Vaz Ferreira, about the use of mathematics as an element of *transcendentalization.* As we said, Geille Castro acknowledges that nothing of the new theory can be understood, deeply, without the use of mathematics. However, this does not mean that the world is mathematical:

> The theory of relativity, will it be the last word of science? Will not a more perfect one appear later? Here is a very important point to highlight, which is the impossibility, for any theory, to express itself, through mathematical notation, in a rigorous form. Fabre, in his book, notes that physicists, to express the laws of nature, which are unchanging, are forced to use mathematical language; they therefore make use of an instrument, like mathematics,

[27] Geille Castro (1924a, 20).

> completely artificial, to express things that are real, natural, immovable. Mathematics is a resource created by the intellect of man and, to demand that it strictly conforms to physical phenomena, is to ask for something completely impossible (Geille Castro 1924a, 143).

The second aspect has to do with the type of training that the new science (mainly physics) seemed to demand from academics and scientists of the twentieth century. Geille Castro speaks on a topic not only global—remember that at that time Max Weber gives his talk on scientific vocation—but also present in his own Faculty of Engineering: scientific specialization. If on the one hand, his work as an engineer but his passion for mathematics and physics, seem to say a lot about the requirements of specialization, Geille Castro advocates for a training that does not isolate scientists from other fields, that allows to think about the problems and theories of each discipline in dialogue and coherence with the other sciences:

> The theory of relativity presents another guarantee: having been made by men of science who were not specialists, but polytechnic. The scholars who intervened in the formation of the doctrine, possessed completely general knowledge; they covered, not only Physics, but Mechanics, Mathematics, Astronomy, and all related sciences. Now, it is known that the result of specialization is, in the vast majority of cases, if not all, negative. And I am not referring exclusively to these of pure science; it is a general phenomenon. Indeed, the specialist sees only his specialty and is exposed to great setbacks for not knowing in depth the other sciences; his criterion is not broad, but limited. Thus, it has been seen that many scientific theories have failed loudly because their authors, being soaked in their specialty and ignoring the others, established doctrines without relation to the other sciences, and that, for that reason, they could not be sustained for a long time.[28]

It is not exaggerated to see how, from that student who publishes mathematical analyses on classic problems of the discipline, by 1923 we have a serious professor, deep in his reflections and very attentive to the scientific advances of his time. But it is also worth showing that these first advances in the theories of modern physics, were still uttered by engineers who did not leave aside their profession or condition as such in the country. And this will not be a minor aspect in the discussion about the value that the political system gave to formations more associated with hard sciences. A year later, from the other organ where scientific knowledge was cultivated in the country, the Polytechnic Association of Uruguay, Geille Castro reflected on the position of engineers in the country.

> In politics, for example, the engineer shines by his absence. In parliament there are a multitude of lawyers, many doctors, but there are no engineers, and that is the cause of more than one mistake, since in all those questions—raised in that room—that have to do with our profession, the authoritative opinion, the experienced advice is missing.[29]

Geille Castro outlines some answers to this problem of the country, not looking for responsibility outside, but focusing on what has become the training of the engineer. The set of these answers points to the problem that the training of the engineer leads him to social isolation, to the development of a character markedly associated with

---

[28] Geille Castro (1924a, 142–143).

[29] Geille Castro (1924a, b, 1–2).

projects and their exclusively technical analysis, and to a form of education focused exclusively on technical aspects.

> We believe that, in addition to the scientific essence and somewhat arid nature of our studies must contribute to this, this cult of the economy must produce a transformation of character. The notion of beauty, the taste for the spiritual, for the superfluous, is somewhat lost, forgetting that all this is precisely the food of the intellect. One becomes sparing in words, sullen in manners, practical in spirit, always looking for the industrial or economic side of every question.[30]

4.3 Enrique Loedel Palumbo, Physics and Physics Education

The other engineering student recognized for his contributions to the development of physics in the region was Enrique Loedel Palumbo (1901–1962). Fortunately, studies on Loedel Palumbo have been much more extensive than those of other early physicists in the region.[31] Without a doubt, his case has been of great relevance for a simple fact: he was the first engineer who decidedly devoted himself to being a professional physicist, abandoning any purpose to develop as an engineer, and showing solvency for his studies. As a counterpart, it should be clarified that the training of Loedel Palumbo does not take place in Montevideo but in what surely was the most developed institution in the field of physics in Latin America, at the beginning of the twentieth century, the Institute of Physics of the UNLP, in Argentina.[32] Specifically, his training in Uruguay would have been until he completed his higher studies and, influenced by Octavio Hansen, he traveled to La Plata to train under the tutelage of the German physicist Richard Gans.[33] At nineteen he published his first book, *New concepts and applications on some points of physics*, a publication that brings together the series of works that until then he had written on the problems of physics. In 1922 an article of his about *The Fermat principle* appears in the monthly student newspaper *Renaissance*, which gives an account of the type of research in which Loedel Palumbo was involved in Argentina, guided by

[30] Geille Castro (1924a, b, 2).

[31] Gangui and Ortiz (2020), Von Reichenbach and Andrini (2015), and Monteiro (2022). https://fisicamartin.blogspot.com/2022/06/la-familia-del-primer-fisico-uruguayo.html. Accessed 26 February 2024.

[32] Gangui and Ortiz (2022).

[33] Richard Gans arrives in Argentina to take over the direction of the Physics Institute after the death of Emil Bose in 1911. Being one of the main physicists working on the new quantum from very early on, he emigrates to these lands due to his Jewish condition and the difficulties he had in finding employment in an increasingly anti-Semitic Germany. Prior to his arrival, two physicists with significant careers temporarily replaced Bose: Jakob Johann Laub and Konrad Simons. The first of them, Laub, was an assistant to Einstein in his years in Bern, Switzerland, working on the new special theory of relativity. The second, Simons, was the one who assumed the direction of the institute immediately after the death of Bose and until the arrival of Gans in 1914. To these physicists we must add the arrival in 1909 of the mathematician Paul Frank, which clearly shows the importance of the Physics Institute in La Plata.

Gans. There we also note the presence of one of the major concerns of Loedel Palumbo throughout his life, that is, the way in which to introduce the teaching of physics in a friendly and accessible way:

> It is curious that this principle of such importance and transcendence in the study of Geometric Optics is not treated in elementary physics books despite its easy accessibility and its very fruitful projections.
>
> (...) Therefore, given that the Fermat principle is so important for the understanding of Nature, as it contributes to the Scientific as well as Philosophical ideal of the unitary explanation of everything and given also how accessible its understanding is, we return to insist on what we said at the beginning, that the principle we study should be introduced as soon as possible in the teaching of elementary Physics.[34]

Loedel Palumbo obtained his Doctorate in Physics in 1925, with the thesis "Determination of the optical constants of sucrose by the investigation of refraction, of the depolarization of Tyndall light and of circular polarization in aqueous solutions of such substance." Two years earlier, in the Faculty of Humanities of the same University, he graduated as a professor of Physics. Research in physics and the teaching of it (for primary and secondary levels), are the topics that will divide the academic career of Loedel Palumbo throughout his trajectory. The topic of Loedel Palumbo's doctoral thesis will appear in the German magazine Annalen der Physik[35] and in the journal of the Institute of Physics, created by Gans himself, Contribution to the Study of the Physical and Mathematical Sciences.[36] Obviously, this is the work of a distinguished student, who will not only become increasingly relevant for his publications in foreign journals, but who will also have the "myth" of being the one who asked the most relevant question to Einstein during his visit to Argentina.

The National Academy of Exact, Physical and Natural Sciences of Buenos Aires was able to organize, within the long but agitated stay of Einstein in Argentina, a meeting dedicated to the theory of relativity, where authorities, and young engineering students like Teófilo Isnardi and Félix Aguilar participated. Loedel Palumbo, although he did not yet have his doctorate, was also invited to the meeting. The research by Gangui and Ortiz on the episode deserves its literal transcription:

> Both the questions posed to Einstein, and his answers, were published in the *Annals* of that Academy, which were then incorporated into the *Annals of the Scientific Society Argentina*. Although the questions from Aguilar, Collo, Damianovich, Isnardi and Loyarte reveal knowledge of this theory, those from Loedel Palumbo, very specific, suggest an interlocutor who at that moment was trying to research in the field of the theory of relativity.
>
> Elsewhere, we have indicated that during some of his trips Einstein kept a "Personal Diary", in which he poured out his impressions. Although this Diary cannot be taken as a completely faithful reflection of his activities, as there are omissions, additions and assumed names in it, his impressions of that meeting were not flattering. On April 16, 1925, Einstein wrote on page 23 of his Diary: "One is asked very silly [inserted: scientific] questions, so it was difficult to remain serious".

---

[34] Palumbo (1922, 5–6).

[35] Palumbo (1926a).

[36] Palumbo (1926b).

> If the official minutes of that meeting, incomplete, could be taken as a reliable reference, Einstein would have found something more interesting in the question posed by the young Loedel Palumbo, which made the visitor stop to answer it, revealing some interest.
>
> The question from Loedel Palumbo was as follows: "Is it possible to find a representation of the two-dimensional space-time surface in a three-dimensional Euclidean space?". The problem, supposedly, had led this young physicist to formulate a system of non-linear partial differential equations that he had not yet managed to solve. According to that same publication, "Professor Einstein answers that [these equations] have not been solved, and that the problem of investigating the form of the space-time surface would be very interesting."[37]

Beyond whether Loedel Palumbo received or not "the blessing" of the famous physicist, the interesting thing is what happened next. The question was not trying to impress Einstein; in a way it was intimately linked to the thesis that Loedel would defend that same year, to the idea of finding a geometric structure in the molecules of sucrose, to the study on geometric optics in Fermat's principle, and, at that time, to the concern to find the geometry of the concept of two-dimensional space-time, which is a simpler and more didactic way of interpreting the space-time events that are understood in Minkowski's diagrams (this geometric novelty was known as the Loedel Palumbo Diagram, whose novelty lies in representing on the same scale the variables of space and time). This solution would appear in the magazine *Physikalische Zeitschrift*.[38] It was an original problem within the theory of relativity, as well as the first original scientific article from Latin America to the theoretical construction of the many open questions left by the theory. But Loedel Palumbo, by then, had built something totally unusual in the Latin American context: to build a line of work of his own that we could summarize as the search for the geometrization of physical phenomena, through mathematics. This line of research continued in 1928 in Berlin, when Loedel Palumbo obtained a scholarship to study in Berlin with prominent physicists like Max Planck and Ernest Schröedinger.

As was mentioned, the scientific production of the young Loedel Palumbo was accompanied by the concern for the teaching of physics and the philosophical problems of science. Loedel Palumbo also ventured into associations and spaces where the concern for the History of Science and the Philosophy of Science accompanied the challenges of the world of physics. On the one hand, during his stay in Berlin he met Hans Reichenbach-and the members of the Berlin Circle-and in 1935 he participated in the first congress of *Philosphie Scientifique* developed in France, where the main exponents of the new scientific philosophy participated. In a CV from the Institute of La Plata from 1940, Loedel Palumbo accounts for what those years meant in his philosophical transformation:

> I am of a passionate temperament; at some stages of my life I studied philosophy fervently; in my youth I was a Cartesian rationalist; later a skeptic, and at the moment the consistent empiricism of the Vienna Circle has my sympathies.[39]

---

[37] Gangui and Ortiz (2020, 829–830).

[38] Palumbo (1926c).

[39] Palumbo (1940).

On the other hand, in Argentina, the Spanish mathematician Julio Rey Pastor and the Italian chemist Aldo Mieli—both with strong links to the Academy of History of Sciences in Paris-created, in 1933, the already mentioned Argentine Group of History of Science, where Eduardo García de Zúñiga participated among the few select members.[40] Loedel Palumbo joined that group, and much of his academic production took a turn from those links with the humanities. In 1936, probably influenced by that philosophical contact with the representatives of logical empiricism, he wrote a work titled *Logic and Metaphysics (An introduction to the study of the problem of causality)*. The book, of few pages, is the product of a lecture given by Loedel at the Philosophical Center of the UNLP, to an audience composed mainly of professors of the humanities, mainly philosophers. There, without rhetorical detours, he firmly attacks Bergsonian philosophy (the one that influenced strongly the philosophical reception of the theory of relativity in Latin America) also demonstrating his knowledge and approach to the program of the new scientific philosophy.

> (...) my exposition will not be the cold account of what others think, since, although there is no originality in it, I have lent it the warmth of what I have lived, for which, what I express there, I feel as my own.
>
> Thus, and given that I will uphold the thesis that all metaphysical propositions lack logical sense, my style will seem to those who uphold the contrary thesis, and who here constitute almost unanimity, excessively polemical and perhaps even hurtful. However, I want to warn, that I have not had at any time the slightest intention of hurting susceptibilities, nor has it been my intention, to belittle the activities of those who dedicate to metaphysics, which, if it is to be assigned any place, according to my criteria, that place would be, between music and poetry, but not below or above science, as a base or crowning of the same.
>
> I can feel a great admiration for Bergson and for the power suggestive and suggestive of his metaphors, without that preventing me to affirm that in all his metaphysical propositions, there is as much of knowledge, or logical sense, as in the music of Mozart or of Beethoven.[41]

Far from considering this passage in a pejorative way, it is convenient to understand it under the life of a physicist in aid of a rebirth philosophy that, precisely, could provide the necessary tools to understand the current state of his science. The place of art did not mean a precinct of lesser category or lesser pedigree, for Loedel Palumbo it was simply another area of human life, which due to its characteristics could not be compatible with that of science. Perhaps a more illustrative way to identify this position of Loedel, which was the version of a Latin American reception of logical empiricism—whose nature was much more complex than the mere reduction to the anti-metaphysical thesis, is to know some of the stanzas of a book of poetry that he publishes a few years before, in 1934, under the title" Verses of a physicist". There are suggestive titles like "Galileo", "Newton", "To Niels Bohr",

[40] Martínez (2023).

[41] Palumbo (1936, 6–7).

"Die Unwahrheit" (the untruth), "Metaphysicians...". It is worth showing an example of his prose that is always handled with humor, in the poem "The value of science":

> Where is your value?—In my beauty,
> in the joy and delight that I provoke,
> in that lies all my greatness
> for the truth, I know well, is worth little.
> But your daughter, the technique, is grandiose;
> Who is there that does not marvel at her?
> My daughter or my stepdaughter is not much,
> For she has not managed to make man happy.
> Does not your end consist in the truth,
> The truth to whom you, validity deny?
> In that I put only my cruelty,
> I have the value of Greek tragedies![42]

In any case, what interested Palumbo was to make the cultural and academic environment of his region understand, is that science not only offers solutions to the problems of Nature, but that it elaborates concepts to attend to these solutions. And this is the greatest aspect of the tension with Metaphysics, as science goes building its concepts on the ruins of the old philosophy-which resists to refound its concepts on its scientific uses:

> All the confusions and misinterpretations of the theory of relativity, all the nonsense of pseudo-philosophical discussions referring to it, come from the ignorance of the right that science has to give itself the definitions of the concepts it handles.[43]

It is interesting to see that his contact with Reichenbach was not only epistolary. In his book *The rise of scientific philosophy*, from 1951, the Berlin philosopher highlights this same obstacle for the development of scientific philosophy:

> The history of speculative philosophy is the story of the errors of men who asked questions they were unable to answer; the answers which they nonetheless gave can be explained only from psychological motives. The history of scientific philosophy is the story of the development of problems. Problems are solved not through vague generalities, or picturesque descriptions of the relation between man and the world, but through technical work.[44]

In 1940, Loedel Palumbo and Salvador De Luca[45] they publish a treaty of *Cosmography*[46] which is the first in a series of works aimed at teaching physics. There begins to be evidence of a methodology in this type of work, the idea of accompanying the teaching of physics with a story that is presented in such a way

[42] Palumbo (1934, 18).

[43] Palumbo (1936, 14).

[44] Reichenbach (1959, 117).

[45] Salvador De Luca was a professor of cosmography and mathematics at the National College of the UNLP and a professor of mathematics at the Girls' Secondary School of the same University.

[46] Palumbo and De Luca (1940).

that it explains the current situation of cosmography, in a way in which concepts and methods used in the discipline are known and reflected upon, framing these knowledge in broader thought systems (much space is dedicated to the cosmography of Kant, Laplace and the new relativity), thus giving a generalist understanding of the corpus of knowledge, without neglecting the more technical aspects (especially the mathematization, measurement techniques and the detail of the experimentation).

In 1941 he produces a seminal work in terms of the teaching of the discipline, called *Elementary Physics*. As Cecilia Reichenbach notes,[47] it was the first textbook that managed to replace an old work by Teobaldo Ricaldoni, former director of the Institute at its inception. The work is of enormous exhaustiveness, sometimes confusing with treatises on research methodologies or encyclopedias of physical terms. It is worth noting that Loedel Palumbo was never a full-time professor dedicated to research; parallel to his position at the Physics Institute, Loedel held positions in the Faculty of Humanities (in the education of physical-mathematical sciences), as well as in schools in Buenos Aires. In this sense, this first work he does is aimed at the training in physics for initial and secondary education, with a highly commendable didactic tone. In the prologue of the work, he shows that his writing was duly crafted in dialogue with the natural recipients of this text, who not only sought to know the classic foundations of the discipline, but its most recent findings:

> They also wanted me to present them with a panoramic view of theoretical physics of our days, convincing me, with irrefutable arguments, that they have the right to be told in plain language, without cumbersome mathematical formulas, what are the fundamental ideas of the theory of relativity or those of quantum mechanics. By articles that appear from time to time in the current newspapers, everyone has learned, for example, that the principle of causality has made a crisis in modern physics.[48]

It will be in the 50 s when the most relevant works of Loedel Palumbo are known, an author who by that time was totally devoted to the production of books that transform the teaching of the discipline.[49] In 1949 is published *Teaching of Physics*, within the collection Library of Educational Sciences, of the Kapeluz publishing house. The director of this collection was Alfredo Domingo Calcagno,[50] and his prologue is not a mere flattering text of the work being presented. On the contrary,

---

[47] Reichenbach (1959).

[48] Palumbo (1941, X).

[49] It is interesting to note that during the 1940s, Loedel Palumbo was involved in extremely rich epistemological debates. In particular, in the company of Ernesto Sábato, he had an interesting discussion with physicist Teófilo Isnardi and Spanish mathematician Julio Rey Pastor, about the physical nature of the concept of temperature. This discussion was reflected in the meetings and publications of the Argentine Scientific Society and the Argentine Mathematical Union, between 1945 and 1946.

[50] Alfredo Domingo Calcagno (1891–1962) was an Argentine pedagogue trained in La Plata and with a doctorate from Brussels, who carried out intense political activity at that university—being its rector between 1945 and 1946, and as an ambassador of UNESCO during the mandate of Frondizi, in 1955.

there he gives a full account of the type of perspective that Loedel Palumbo sought to develop with this type of texts. The first of them, was the demand for the specialization of physics and how to solve it when a comprehensive training of the student is intended:

> (...) a fundamental aspect of teaching, in its formative function, is to overcome all the difficulties that the inevitable specialization opposes to the development of culture. If it is necessary to subdivide, it is first essential to unify and ensure that those who are going to operate in a very limited sector of the wide and deep stage of human knowledge, of social life and of constructive action, are interested in the surrounding reality and are capable of appreciating and understanding, in their general lines, the work of those who act in distant parcels of their own.[51]

In a way, what Calcagno recognizes in Loedel Palumbo's work is the relevant role that this kind of texts/manuals will begin to play in the training of future scientists. In them is the key on how to act in the face of the specialist demand in which physics was found. These are concerns that anticipate a decade what Thomas S. Kuhn will develop more theoretically in his book *The Structure of Scientific Revolutions*, about "the invisibility of revolutions."[52] In this passage of the work, Kuhn warns about the role that textbooks play in scientific education, where generally the history of science occupies a purifying place of what the scientific practice has been really. According to Kuhn, textbooks operate as distortions by showing only fragments of the history of science that are placed at the explanatory service of current science. Well, none of this should be said, or at least lightly, about Loedel Palumbo's works related to the practice of Physics. While it is true his In the spirit of showcasing a "cumulative" science, the focus is far from confining history to the sum of crucial experiments to understand current physics; on the contrary, Loedel is concerned with presenting these discoveries in broader frameworks of thought, especially within the structured philosophies or cosmologies of each era. As Calcagno highlights, these texts by Loedel Palumbo take on the complex task of fighting against specialization in search of forming scientists whose technical understanding is not confined to formulas, and always maintains a more comprehensive view of physics in the body of scientific knowledge. And this characteristic seems to have been one of the notable attributes that stood out in the personality of Loedel Palumbo, as the following anecdote presented by Calcagno well attests:

---

[51] Palumbo (1949, XVIII).

[52] "For reasons that are obvious and very functional, scientific textbooks (and too many old histories of science) refer only to the parts of the work of past scientists that can easily be seen as contributions to the statement and solution of the paradigmatic problems of the textbooks. Partly by selection and partly by distortion, the scientists of previous times are implicitly represented as if they had worked on the same set of fixed problems and according to the same set of fixed canons that the most recent revolution in scientific theory and methodology has made appear as scientists. It is not strange that both textbooks and the historical tradition they imply, have to be rewritten immediately after each scientific revolution. And it is not strange that, when rewritten, science appears, once again, largely as cumulative" (Kuhn 2015, 215).

> One afternoon Henriquez Ureña[53] arrives at the rector's office, with his short and hurried step, and tells me, more or less: "Well, here I bring a decoration for a colleague. I have just reviewed these compositions of my boys on the topic we had agreed upon ('Praise of the matter I like the most'). And see: in astonishing majority the students have preferred the new course of Loedel and have written the 'Praise of Physics'. It seemed to me that the rector should also know this." We pleasantly discussed the matter and he left me the compositions and his congratulations for the Physics professor. Henríquez Ureña knew well that, as a rule, high school students, at a time when preferences are not yet defined or vocations oriented, show a preference for the subjects that are taught best; and he, the great teacher, so beloved and respected by all in the College, who took care of each of his courses and each of his students with capacity, dedication and exemplary responsibility, at the same time as the best, considered it a duty to bring me such proof of the teaching efficiency of a wise colleague. What glory it was to work with men of such quality![54]

The book contains a contribution that has been highlighted by the Uruguayan engineer and physicist Walter S. Hill, as "an idea at once simple and brilliant and constitutes the best didactic proof devised to date of the earth's rotation."[55] This experiment appears in a section of the book called *Physics at home*, aimed at students experimenting with the phenomena of physics in their own homes, extending the teaching environment beyond the classrooms.

> I will end this chapter by indicating to you how to construct something like a "microscope of rotations", with which device you will be able to verify, in just a few moments, the rotation movement of our planet. The new proof, which I carried out for the first time in the middle of the year 1946, when I had already begun to write this book, is based on the principle of the conservation of rotational momentum.[56]

Finally, in 1955 appears what will be-probably-the most important book of Loedel Palumbo: *Relativistic Physics*. Designed to pay tribute to the special theory of 1905, on its fiftieth anniversary, it became a general tribute to the figure of Einstein, who died that same year. It can be established that it is his most important book because it condenses his two great intellectual projects: the scientific work in the field of relativity that he carried out in his early years as a Doctor in Physics, along with his vocation for thinking and proposing new ways for the teaching of physics. The tone of the book does not differ from those he published in the forties: from a everyday language, without avoiding mathematical challenges, Loedel Palumbo historically reconstructs the most important intellectual development of the twentieth century, the *relativistic physics*. His motivation is simple, bring the Einsteinian theory to general culture, fight against the resistance to believe that this knowledge so

[53] Pedro Henríquez Ureña (1884–1946) was an intellectual born in the Dominican Republic, who emigrated early to the United States, continuing his intellectual activity in Mexico and Argentina. In the latter country, he worked at the National College of La Plata, as well as in the chair of philology at the UBA. He was one of the most present influences in the young generation of Argentine writers, among whom were Ernesto Sábato, Silvina Ocampo and Jorge Luis Borges.

[54] Palumbo (1949, XXXI).

[55] Palumbo (1949, XXVIII).

[56] Palumbo (1949, 251).

relevant to our era can only be restricted to a select group of people. In the article of that same year, already mentioned, where Loedel Palumbo pays a tribute to the figure of Einstein, he said the following:

> (...) of this conceptual revolution, analogous to the Copernican revolution, what can a cultured man of our time say? What does a doctor, a lawyer, an engineer and even a physicist know about it if he is not precisely a specialized theoretical physicist? Within a few months (in June 1955) it will be 50 years since the appearance of the first Einstein paper and... what is taught about the Einsteinian theory in the official courses of secondary and university education? The daily press has known how to be more in tune with the times than the official teaching institutions. Through it, for better or worse, people know something about Einstein, whom they imagine as a strange being who takes out of his retorts strange symbols of an even stranger mathematics and which are then translated into unexpected relations between mass and energy, the weight of light, the curvature of space and the total size of the universe. The theory that appeared in its time to put order in the panorama of the Classical Physics, of an extraordinary conceptual simplicity, is presented, by a strange paradox, in the mind of the common people, as a chaotic tangle totally indecipherable.[57]

Loedel Palumbo's ties with Uruguay always managed to remain. In addition to the relationship he maintained as a correspondent for some magazines, Loedel Palumbo participated in teaching activities of Physics at Udelar due to his connection with Walter S. Hill, who was the central figure of Physics at the university since the forties, as we will see below. In 1948, invited by Hill himself, Loedel Palumbo conducted a course on relativity at the recently created FHyC.[58] On his part, Walter S. Hill was very complimentary about the experiment created by Loedel Palumbo to detect the rotation of the earth. This link between both allows us to see the existence of this Rioplatense connection, and helps us to introduce the figure of Walter S. Hill, as the third character who early on dedicated his efforts to the cultivation of physics in Uruguay.

4.3.1 *Emilia Loedel*

As mentioned, the Loedel family developed professional activities in several professional and scientific areas. Enrique's brother, Eduardo, was a doctor who worked at the Faculty of Medicine. Greater attention deserves his sister Emilia, the first engineer of Uruguay graduated in 1920, along with Juana Pereyra. In 1929 a her essay appears, relevant for its early appearance in the context of the physical discussions in the country, and for being from a female author, of whom there are very few records in these times referred to this type of discussions. The text is titled *Gravitation*, and it breaks down into two sections: (a) *history of the development of the conceptions about the nature of gravitation*, and, (b) *gravitation as a manifestation of the non-Euclidean nature of space.* Despite being a small text, it condenses

[57] Palumbo (1955, 73).

[58] Hill (1948).

with an uncommon clarity for the time, the difficulties faced by scientific thought when pursued by metaphysics. In her brief history of the concept of gravity, Emilia Loedel traces the substantive question about the dividing line between science and metaphysics, analyzing the contributions of Galileo, Kepler, and Newton. This type of analysis emerges, according to her, with Comte's positivism:

> Comte's positivism marked as the dividing line between Science and Metaphysics, the same limit that between the "how?" and the "why?". Thus, Science should only be concerned with describing **how** phenomena occur, without even inquiring about the causes that produce them, that is, the " why" insinuating, that when presented to man, they constitute a challenge, a challenge or a hope; flattering and happy hope that is enough to multiply his stimuli, until committing him to the quixotic adventure of dispelling shadows, to reach the presumed light, through the silent harmony that is felt rumbling in the outside world... The history of Science teaches us, on the contrary, that the alleged dividing line between the **how** and the **why**, is far from being well defined and therefore clearly marking the border of scientific knowledge; thus it is observed that as Science advances, it follows it with the same step, in the same way that the blue limit of the horizon remains, constantly, at the same distance from the navigator who pursues it.[59]

Even more interesting is the second section, because it allows us confirm the effect that the changes made in the training of engineers, carried out by Eduardo García de Zúñiga-especially, in geometry-, caused in many students who learned about these new plans. In the case of Emilia Loedel and her article on gravitation, it shows how to understand that the physical possibility of new geometries that explain gravitational movements, are derived from theoretically harboring a plurality of geometries. It will be observed that there are important qualitative differences in the approaches of these young engineers, knowledgeable about the most technical issues of non-Euclidean geometry-which serves as the foundation for the general theory of relativity-, compared to the approaches that will give old professors like Reyes Thevenet or Williman.

> We believe, driven by our intuition, that our physical space has three dimensions and that it is Euclidean; however, who assures us that this world of three dimensions is not curved in the sense of a fourth dimension, that we cannot appreciate, and that, if we could make measurements of extremely large triangles in our Universe, we would not find that the sum of its three angles, is greater or lesser than two right angles? (...) For now, from the point of view logical, there is absolutely no reason to prefer between the Euclidean geometry and any other, and we will see later that previous certain considerations and with respect to the nature of time, **Gravitation is nothing more than a manifestation of the non-Euclidean nature of our space**.[60]

Finally, in the same tone of reflection as her brother Enrique, Emilia clearly dispels that the path that the challenges that the metaphysical questions present, are not foreign to science but that constitute its final orientation.

> We believe we have shown that the man of science, far from disregarding the " whys" we referred to at the beginning of these considerations, is stimulated by them, and his difference with the metaphysician would consist in that when climbing the rugged slope of the

[59] Loedel (1929, 14).

[60] Loedel (1929, 16–17).

> Unknown, he does so step by step, giving each step taken the value it has, for the effort it signifies and for the new perspectives that it opens to the spirit to be able to contemplate the panorama from a greater height, since it is risky or impossible to ascend this unlimited scale, without climbing decisively and patiently, rung by rung.
>
> Is this a real ascent or just an appearance? However, from the subjective point of view, there is no doubt that this constant hustle of the struggle to which man is subjected, in trying to scrutinize the secrets of Nature, produces satisfactions that greatly compensate for the effort made, giving him at least the illusion that he advances and advances incessantly.[61]

4.4 Walter Scott Hill and the creation of the Institute[62]

Walter S. Hill trained as an engineer between 1922 and 1927, making a career as a student that allowed him to obtain a distinction among the most outstanding of his generation. Just a few years after obtaining his degree he was awarded the fourth incentive prize for his innovative work titled *Mathematical theory of the pendulum transformer*, presented in an academic competition within the faculty. He began his connection with teaching and especially with physics in 1931 when he obtained through competition by opposition and merits the chairs of Optics, Hydrostatics and Technical Physics. In 1933, he taught the course on Machines, specifically for Industrial Engineering students, as a free chair. His career continued with key appointments, such as full professor of Atomic Physics at the Institute of Higher Studies (IES) in Montevideo in 1934 and director of the Physico-Mathematical Section of the same institute in 1936. That same year he was incorporated as a founding member of the Argentine Mathematical Union.

He began his scientific career with the publication of the article *Knapen Effect* in August 1926, a technical dissemination work published in the *Engineering Journal*, organ of the Center for Engineering Students. A year later, in 1927, Hill presented a study titled: *Mathematical theory of the pendulum transformer*, for which he was awarded the incentive prize of the Faculty of Engineering. That same year he published his *Physics notes: Method of analogies*, in the September 1927 and August 1928 issues of the *Engineering Journal* of Montevideo. He also publishes in the *Engineering Bulletin* on the transpiration of masonry in 1931, an article of his appeared in the *Engineering Journal*, titled "Mechanical equivalence of light", his first work of theoretical physics.

In 1932, Hill continued with his scientific contribution by publishing "*Electric Television*" in the *Engineering Journal* official of the Association of Engineers of Uruguay, and presented "*Gauging using the pendulum*", a work on applied hydrodynamics, published in March of the same year. Other publications from his early

[61] Loedel (1929, 19).

[62] The information in this passage constitutes a summary of the work carried out in previous research. For complete information see Queijo Olano (2023) and Queijo Olano and Videira (2022a, b).

period include studies on the *Transmission of heat by radiation* (1932–1934), *Thermodynamics Course* (1932), *Note on the dynamic equations and thermodynamics of fluids* (1933), and theoretical-experimental works like *Stabilization by piezoelectric quartz* (1933).

Throughout his scientific activity, Hill showed a clear interest both in the theoretical foundations of physics and in its practical applications. This theoretical approach, focused on the conceptual bases of physics, is reflected in several of his later publications. Among them is the "*Note on the Equilibrium of Fluids*" (1936), published in the *Engineering Journal*, where he addresses the mathematical physics of continuous media. In 1937, Hill presented the class notes titled "*Thermodynamics and Fluid Mechanics*", a teaching work developed to opt for the chair of Mathematical Physics at the University of Montevideo. That same year, he participated in the III International Physics Congress with the work "*A definition of real fluids*", in which he expanded his research on the thermodynamics of fluids. In addition, he presented the work "*Heat transmission by conductivity. Convection problem*" (1937), also as part of his candidacy for the Chair of Mathematical Physics, where he explored the physical mechanisms of heat transmission, a central topic in thermal physics and its applications.

Likewise, Walter S. Hill focused on researching technical problems with direct practical applications, as seen in several of his published works in the *Engineering Journal*. In 1936, he presented "*Thermoelectric Generator*", a theoretical-experimental work on energy generation through thermoelectric principles, and "*Concentration of concretes based on the compactness of the mixture*", where he applies principles of physics and mechanics to the optimization of concrete, a key technique in civil engineering. In 1939, he collaborated with architect Julio Vilamajó on the article "Faculty of Engineering, its building under construction", also published in the *Engineering Journal*, where they analyzed architectural and structural aspects of the Faculty of Engineering in Montevideo. In 1940, Hill published the study "*Absorption of Fluorescence Light*", a theoretical-experimental work presented to the Academy of Sciences in Rio de Janeiro, which addresses the interaction of light with fluorescent materials. Finally, in January 1941, his "Notes on the Chair of Electrical Conductors" were published in the *Electrotechnical Journal* of Buenos Aires (volume 26), where he explores the principles and applications of electrical conductors, a fundamental topic in electrical engineering. These investigations not only present a solid theoretical basis, but also explore the practical implications of these phenomena.

These initial academic essays account for two characteristic attributes of Hill as an engineer and physicist: a great ease for developing mathematical calculations—a practice that would benefit him in his engineering activity, but also within physics—probably derived from the transformations made in the study plans of the career-in its emphasis on mathematics-, as was previously commented. But on the other hand, Walter S. Hill if we observe his development as a physicist, we will see a constant restless movement that leads him to land on all kinds of matters without delving into them deeply. Thus, we will see his active participation in engineering problems, theoretical physics issues, practices in experimental physics and even to venture into the relations of physics with nuclear medicine or biology.

In 1937, Walter S. Hill accessed by competition to the Chair of Physical-Mathematical Sciences at the IES and a year and a half later presented communications at the Third South American Chemistry Congress.[63] In those years, the Faculty of Engineering laid the foundations for its move to a new building, in whose construction Hill would also play an important role.[64]

The thirties passed with much regional participation of Hill in congresses and seminars: the production on some topics of academic interest, which was presented at congresses in Buenos Aires, Rio de Janeiro and Washington,[65] and the obtaining of resources to buy materials and equipment for the Physics Laboratory, of which he was already head in 1937. These data are not merely anecdotal. They are the first successful efforts to establish a space for the practice of theoretical physics in the country, whose instruments were not placed at the service of the training of engineers and allowed the development of the discipline (which we will see more in detail in the next chapter). It is worth establishing what this development of the Institute and physics consisted of, manifest in Hill's letters and writings: a tenacious idea of the laboratory as the central reference of all scientific practice. This is a vision—at least—original, for two elements: on the one hand, because it removes the academic chair as the space of validation of teaching knowledge and moves it towards the laboratory, the space of practice. This movement is extremely interesting if we take into account some ideas that he will express later in his life, about the vision he had of the chair in relation to the laboratory. A second element is the way of developing physics: it was undeniable for him that the divisions between theoretical physics and experimental physics—which in those years were accentuated in the most relevant research centers in the world—did not represent a real problem. According to him, there could be no real teaching of physics without a laboratory. However, from the writing of popular science texts on the modern advances of the discipline, Hill contributed to the regional development of the discipline.

His book published by the University of Litoral was recognized as one of the first contributions in the construction of the community of physicists in the city of Rosario, Argentina. That book was called *Nuclear Combustion*. The historians of physics Galles and Rivarola, highlighted the relevance of this work in the following way:

---

[63] At that congress he presented the paper "On a definition of real fluids" (1937).

[64] For a study of Walter S. Hill's role in the work, consult the thesis by Scheps Grandal (2008).

[65] The papers he presented, varied in their themes, were: in 1935, the lecture "Compatibility of waves and particles" and the courses "Aesthetic Definition of Real Fluid"; "General Equations of Internal, Surface and Linear or Boundary Equilibrium"; "General Thermodynamics of Problem"; "Introduction of Internal, Surface and Boundary State Equations", at the Seminar of Mathematics and Exact Sciences of Buenos Aires; in 1939, at the First South American Engineering Congress (Uruguay) and in the *Engineering Review of Buenos Aires* (Argentina); "Flat slabs of reinforced concrete with crossed reinforcements"; in 1940, at the Washington Congress (United States), "A research of equilibrium of fluids"; in the *Annals of the Brazilian Academy of Sciences*; "Absorption of fluorescence light by the fluorescent liquid"; and in the *Engineering Review* (Uruguay), "Introduction of obstacles in the courses of rivers".

> There was an interesting precedent of considerable importance in Rosario on the issue of nuclear energy that is not often mentioned. As is well known, a censorship was established in the United States from April 1940 on the publication of research on uranium fission; something similar happened in all the warring powers. Neutral countries had more freedom in this regard. The state of nuclear physics was analyzed in an article published in the Technical and Scientific Series of the Faculty of Mathematical Sciences titled "Nuclear Combustion", prepared by engineer Walter. S Hill, director of the Physics Institute of the Faculty of Engineering and Minor Branches of Montevideo. Hill pointed out the importance of Hahn and Strassmann's discovery, after which it was understood that it was possible to achieve the chain reaction, releasing a large amount of energy in a short space of time, which "makes the process have the characteristics of an explosion". Hill mentions the work of Francis Perrin in which the critical mass is estimated at about twelve tons of uranium. Undoubtedly, Hill's article can be considered pioneering in the dissemination in our environment of this subject.[66]

In his final reflections, Walter S. Hill warns about the possible consequences of the uses of nuclear energy, foreseeing the problems that the discovery of nuclear fission will cause in the world. It is worth noting that Walter S. Hill, in this book, not only accounts for the details involved in this discovery, the calculations on the impact of the amount of energy that is released in the fission of the uranium atom; but also shows a precise knowledge about the most finished details of the process of this discovery. It is both a complete and didactic work, a conceptual approach to the discovery of nuclear energy and the necessary equipment for its use, and warns about the ethical dimensions of this type of research in scientific life:

> We are, undoubtedly, in a realm of fantasies, but not of *crazy fantasies,* Never before has there been such reasonable discussion of these problems, which coming from the most abstract positions of metaphysics and chimera, have today become concrete in laboratory experiences.
>
> The vision of such wonderful possibilities leads, perhaps, a little naively and impulsively, to predict the good that will derive from them for man and for future civilization.
>
> But let's be cautious and curb that optimism a bit.
>
> One wonders first, if, once such a process is launched, unleashing hidden and powerful forces of nature—an irreparable catastrophe would not ensue. The progressive or explosive, but fatal destruction of the world we inhabit, through a contagion phenomenon, a kind of inextinguishable nuclear fire. According to this dire hypothesis, our planet would burn, in a terrible irreversible evolution, breaking its millennial stability, by a mechanism as simple as that which allows unraveling a fabric by pulling the end of one of its threads.
>
> But this is not the only danger we expose ourselves to in our desire to steal "its fire" from the nucleus, nor perhaps, the worst. Our generation already has finished experience on the abuse that man usually makes of the noblest products of his mind. The current war is also a product of technique, or rather, of the perversion of technique, organized to destroy and subjugate.

[66] Galles and Rivarola (2012, 160–161).

> A nightmare vision, opposes the first, and makes one doubt science considered as a factor of human progress: the problem of the morality of science is raised, to which Professor Cortés Plá refers in "Some aspects of Modern Physics".
>
> How is it possible that the fruit of the quiet, generous and disinterested work of the researcher, can turn into an element of death and destruction, turning against men?
>
> How is it possible that such an irrational use is made of inventions intended, in the intention of the wise or the inventor, to improve the life of man?
>
> Should we renounce science and give up research?
>
> These and many other questions of this kind, surely we have all formulated many times.
>
> The cause of this paradoxical evil is not in science; it is in men.
>
> Science, driven by an elite, has always been at the forefront of peoples and its intention has always been generous and high.[67]

In their differences and diversity of trajectories, Geille Castro, Loedel Palumbo and Hill make up names of a generation that shows some particular characteristics. The three, from the Faculty of Engineering, find a vocational development in the field of physics, although this does not always mean the total abandonment of the practice of engineering (as it was in Geille Castro and Walter S. Hill). This vocation had a very clear condition of possibility, which-we understand-is linked to the transformations that the mathematics study plans propose to the training of engineers, changes driven mainly by Eduardo García de Zúñiga. The three, in turn, experienced the impact of the new physics-especially after the publication of the general theory of relativity-very closely: Geille Castro gives the first course and publishes the first work in the country on the Theory of Relativity; Loedel Palumbo writes the first Latin American scientific work on one of the issues to be answered that the theory left; and Walter S. Hill-present in the student delegation that receives Einstein in Uruguay-seeks to develop a theoretical physics when he assumes the highest responsibilities in the Physics Laboratory first, and in the Physics Institute, later.

It is these three names who mainly conceive the possibility of a scientific development of physics in the country, within the Faculty of Engineering. The general theory of relativity not only aroused their admiration and astonishment, it was an impact that exceeded the restricted scope of the very modest physics that was being done up to that time. This theory and Einstein's visit seem to have opened a schism in the academic community that until then no one would have raised (with the exception of García de Zúñiga or Kayel): a mark that begins to separate the way in which to understand the development of science at Udelar, and that divides an old humanist encyclopedic tradition from a young group that defends the professionalization of scientific activity. But as in other contexts, there was philosophical skepticism in Uruguay towards the new physics proposed by Einstein. In the next passage we will see some of these resistances.

[67] Hill (1941, 68–69).

4.5 Cultural Resistance to the Theory of Relativity

As noted earlier, the tradition in mathematics, geometry and algebra should not be considered as initiated by García de Zúñiga. Rather, the works of Nicolas Piaggio were undoubtedly the impetus for some of his students to continue his legacy. Such was the case of Alberto Reyes Thevenet, professor of Cosmography at Udelar and one of those responsible for the installation of the first Astronomical Observatory of the University, dependent on the Secondary Section. In 1920, in his speech for the application for the chair of Cosmography—which belonged to Piaggio and to whom Reyes Thevenet dedicates words of praise-, the candidate makes a positivist journey through the history of Astronomy, using the stages of human evolutionary progress as Comte and Spencer did, from a formation of mystical thought to its positive stage.

But this positive stage of Astronomy, for Reyes Thevenet, was achieved and culminated in the modern era with the work of Kepler, Galileo and Newton. It is interesting to show not only how this is argued, but the bibliographic support on which this narrative is built:

> And finally comes the profound influence that astronomy exerts in the modern era on all science in general. In the purely speculative order, nothing is comparable to the revolution caused by the discovery of Copernicus of the true heliocentric system[68]; and the new and daring conception of the system of the world, which emerges as a result of his brilliant "meditations" destroyed, in one fell swoop, the entire scholastic block of the Aristotelian-medieval world. The series of conquests that for the glory of human thought, the astronomers of the sixteenth and seventeenth centuries made, and which form, together, according to the beautiful expression of Sageret,[69] a harmonious building in which beauty is reflected at the same time on the Universe and the spirit. It is a complete flight in essence, in conception, in the method of ancient science that takes place as a result of the discoveries of Kepler, of Galileo and of Newton in Astronomy, in Physics, in Mechanics and in all natural sciences. The idea of relation, of constant relations between variable magnitudes, which modern science adopts as the basis of its analytical conception, comes to us originally from Kepler, concretized, "by almost divine calculations", in an astronomical law that Bergson[70] classifies "as the very type of scientific knowledge as we understand it today". And the same can be said of Galileo's law on the constant acceleration of falling bodies. And the same of Newton's law "that has agreed in a formula that fits in the palm of a hand, all the movements of the Universe."[71]

The positivist imprint of Reyes Thevenet's text anticipates his few and disdainful words that he will utter about the general theory of relativity. His training, in the

[68] Reyes Thevenet cites here the work of Ferdinand Hoefer, from 1873, *History of astronomy from its origins to our days* (Paris: Libraire Hachette et Cia.), a German-French doctor, who philosophically wrote about some episodes in the history of science.

[69] *The system of the world. From the Chaldeans to Newton* (Paris: Libraire Félix Alcan, 1913), by Jules Sageret, another writer of the history of science and French philosophy, contemporary of Émile Meyerson and Abel Rey.

[70] Of the influence of Bergson in the Rio de la Plata scientific formation, the work that Reyes Thevenet cites in this passage should be highlighted: *Creative Evolution*, from 1917.

[71] Reyes Thevenet (1920a, 28). This final passage represents a quote from the work of another French philosopher, Fouillée (1891).

strict sense, is consistent with the philosophical tradition from which certain areas of knowledge, such as physics, are still not free. This is reflected in the selection of works that support the perspective proposed: histories made by writers and philosophers, who place the end of the evolution of astronomy in the figure of Newton, and who represent a particular style of French science historiography, more popular than professional (perhaps with the exception of Jules Sageret). Therefore, it should not surprise what is mentioned next about modern physics:

> And this great conquest of the Law, which is, in the last term, the end fundamental of science, we owe it to Astronomy, affirms the wise Professor of the Sorbonne [referring to Henri Poincaré], and constitutes by itself, the greatness of this science, greater even than the material greatness of the objects it considers. In this sense they are, also, definitive concepts that science has seen incorporated into its bosom by the astronomical spirit, because despite modern attempts to modify them, nothing allows to assure, for now, that it can be reached, one day, to a more exact approximation than the sesquialteral relation (arithmetic mean between 1 and 2) of Kepler's harmonic law, nor that the precise terms of the Newtonian formula can be altered either, by the modern theories of Luminous Gravitation[72] and of "Relativity", despite the fact that these perfectly account for the famous anomaly in the perihelion of Mercury.[73]

Reyes Thevenet publishes, that same year, in an article titled *Astronomy in Science*,[74] where he reiterates the type of references that constituted his dissertation for the chair of Cosmography. Again he mentions nothing of the astronomical revolution that was happening from the confirmations made by Eddington a year earlier, although they were undoubtedly in his knowledge.

Relativity presented to preparatory students of the university written by José Claudio Williman in 1923, is a book designed to teach Einstein's Theory of Relativity to preparatory education students in Uruguay. Williman presents the concepts of relativity in an accessible way for students who are starting in physics, therefore, Williman's focus on the teaching of physics seems to be centered on making the complex concepts of the theory of relativity, such as time dilation, length contraction and mass-energy equivalence, understandable without requiring an advanced level of mathematics. His approach is based on an understanding that he describes in two stages: the formation of a representative image of the idea and the integration of that idea into the framework of already acquired knowledge.

> The purpose of this exposition is not to say more than what has been said about this topic, but to contribute to the clarification of some of its fundamental points, taking into account the preparation of those who are going to read it. From a general point of view, when it comes to making and understanding a theory, its consequences, verifications and the philosophical comments it arouses should be avoided, in its exposition, as much as possible, giving only importance to the essential points. The expansion of a theory must be subsequent to the fact of having understood it and not simultaneous.[75]

---

[72] Here appears the only reference, although in French, of a non-French author, as was the Italian mathematician Federigo Enriques. The work cited by Reyes Thevenet is: *The fundamental concepts of science*. Paris: Flammarion, 1913.

[73] Reyes Thevenet (1920b, 29).

[74] Reyes Thevenet (1920a).

[75] Williman (1923, 3).

In presenting his teaching method, Williman criticizes the science popularization books available in his time, pointing out that many of them fail to meet the necessary conditions for effective teaching. He considers that some books aimed at a non-specialized and heterogeneous audience, tend to cover too many points, which makes it difficult for the reader to extract the essentials. Regarding the books that are too specialized, although he recognizes that the precision of mathematics contributes to the physical sciences, he warns that its use can hide the true aspect of physical phenomena, making their understanding difficult.

> The intervention of Mathematics in the teaching of Physical Sciences has the advantage of leading to more precise results and the disadvantage of hiding the true aspect of the phenomenon. Therefore, when it comes to understanding, the mathematical procedure must be, as much as possible, eliminated. It makes one reason about mathematical signs, and from those signs the images that are needed to think, and, many times, it is assumed the matter is understood, when in reality, only the mathematical formula that contains the laws of the matter has been understood. When it comes, not to understand existing knowledge, but to discover new knowledge, the role of Mathematics is different and superior.[76]

Williman admits that some parts of the theory of relativity are so intricate that they require knowledge of mathematics to be understood with greater depth and precision. However, he argues that it is more valuable to make many understand something even if it is in an approximate way, than to insist on an accuracy that few could follow. He even points out that mathematicians often do not fully understand the scientific discoveries that are expressed in mathematical language, and that while the use of formulas allows precision when elaborating the theory, the same mathematicians who work with the formulas, often do not see beyond them. The interpretation of the formulas and their translation into common language is not always achieved, which underscores the gap between technical knowledge and accessible understanding. From these considerations, Williman, sets out to reconstruct a historical and conceptual base of the main problems and explanations of physics at the end of the nineteenth century and the beginning of the twentieth century. This reconstruction takes him six chapters, in which he focuses on bringing closer the most fundamental aspects such as the debate around the hypothesis of the ether, highlighting the inconsistencies and limitations of classical physics that paved the way for the arrival of relativity.

Thus, in Chaps. 1, 2 and 3 he begins by exposing the implications of two seemingly contradictory astronomical discoveries about the speed of light that raise questions about whether the Earth drags the ether and if there is a speed of the Earth relative to the ether. These questions are linked to the notions of absolute motion and rest and are presented as the origin of modern theories of relativity. Williman discusses how the traditional idea of absolute rest, associated with a reference system at rest, is problematic; showing how classical mechanics assumes that some stars were in absolute rest, serving as reference points for laws such as that of inertia.

---

[76] Williman (1923, 4–5).

> Classical Mechanics assumed some stars were at rest, and, with respect to the system of axes at rest in relation to these stars; called Galileo's reference system, the laws of motion were certain, among which was, in the first place, the law of Inertia. Not being able to accept the absolute rest of the stars presents the problem of looking for another reference system. equivalent to what before was that of Galileo. As it is useless to talk about the center of gravity of the Universe, which, if it existed, would be of unknown position to us, the matter, in these last years, was posed in the following format: *If the Ether does not participate in the movement of bodies, all movement or rest in relation to the Ether will be considered absolute; but if the Ether participates in the movement of bodies, absolute movement and rest do not exist.*[77]

The historical-conceptual discussion made by Williman reflects the state of science at the time of transition, where the existence and function of the ether were being seriously questioned, culminating with Einstein's special theory of relativity, which dispenses with the need to postulate this entity. This dilemma is crucial because it questions the notion of a static and universal ether that permeates all space and serves as an absolute reference for movement. Williman, in this chapter, introduces the Michelson experiment, which is key for understanding one of the most important turning points in pre-relativity physics, and he does so thinking of an audience that might not be familiar with the details. Although the experiment is technically complex, Williman manages to simplify the concept for better understanding and for that he uses analogies; he uses the example of an airplane and the wind where it moves, to clarify why the experiment showed that there could be no "ether wind" (that is, that the Earth does not move through an ether medium). But on the other hand, Williman also shows that the aberration of light and other observed phenomena support, at least in part, the idea that the ether might not be completely discarded.

> The Michelson experiment was done taking all possible precautions, and was repeated at different times of the year and with different orientations of the apparatus, sometimes involving other experimenters, and always with negative results. This result gave reason to Sitter, that is, to the hypothesis according to which the Ether participates in the movement of the Earth. But one could not dispense with the Aberration of Light and its consequences and other phenomena, perfectly proven, and to which we have not referred to not complicate, and that also support the Hypothesis of the immobility of the Ether.[78]

By explaining the experiment in an accessible way, using simple analogies and highlighting the unresolved problems in classical physics, Williman tries to facilitate the understanding of the principles that led to the development of the special theory of relativity. His work, in this sense, seems to fulfill the purpose of being a resource for introducing students to the conceptual and philosophical debates of science at that crucial moment in the history of physics. To enrich the discussion about the ether, he introduces the explanation of the Fitzgerald-Lorentz Hypothesis, formulated independently and almost simultaneously by both scientists at the end of the nineteenth century. This hypothesis, which holds that a body moving through the ether experiences a shortening in the direction of its displacement, was, for

[77] Williman (1923, 8).

[78] Williman (1923, 11).

Williman, an ingenious solution that tried to reconcile the negative results of the Michelson-Morley experiment with the theory of the ether, without completely eliminating it. This proposal reflects how scientists were looking for intermediate explanations before Einstein offered a broader solution with his special theory of relativity. By explaining the Fitzgerald and Lorentz hypothesis, Williman seeks to effectively connect the experimental results with the theoretical hypotheses developed to explain them.

In Chap. 6, through a clear and pedagogical explanation, he presents the fundamental elements of the *Lorentz Transformations* and their relationship with the special theory of relativity highlighting one of the most striking results of these transformations: the idea that time and space are not absolute, but depend on the moving reference system. Throughout the chapter, Williman uses analogies and concrete examples to make more understandable the effects of the Lorentz transformations. A particularly illustrative example is the case of the traveler who moves at high speed and, upon returning to Earth, discovers that his children have aged while he has barely aged. This is one of the most used metaphors to generally understand the idea of a time dilation. In this context, he also underlines the fundamental differences between classical physics and the new interpretation that Lorentz offers, highlighting that classical mechanics is still valid in certain contexts, but is insufficient to describe phenomena in relativistic situations.

In Chaps. 7 and 8 Williman directly addresses the *Einstein's Special Relativity*, considering that this theory arises as a bold extension of Lorentz's ideas, but significantly expands its application by establishing that natural laws are the same for all inertial systems, regardless of their movement. Williman explains the relativity of time and space, using examples such as the simultaneity of events to demonstrate that these notions are not absolute, but depend on the reference system. Although he recognizes the difficulty of fully interpreting the relativity of time, he again criticizes the excessive dependence on mathematical calculations, suggesting that understanding must go beyond abstractions. In addition, he introduces the *General Relativity*, using the analogy of measurement and the use of the concept of reference mollusk, showing how space-time curves and adapts, in contrast to the rectilinear systems of classical mechanics. Gravity is no longer analogous to a force, but the consequence of the curvature of space-time. In addition, the author delves into complex concepts such as the weight of light, a phenomenon that demonstrates how light is affected by gravity, challenging classical notions that linked gravity exclusively with mass. In the explanation he uses examples about the deviation of light in gravitational fields and highlights the importance of experimental checks, pointing out that General relativity requires precise observations at interplanetary scales to be verified.

In conclusion, it provides a recap and reflection on the theory, addressing both its scope and limitations as well as its scientific value in the broader context of physics.

> The Michelson experiment can be interpreted in a third way, different from Lorentz's and Einstein's, which consists of assuming what we might call two ethers or two types of ether, only one of which accompanies the movement of bodies in the form of a limited extension

atmosphere. As we say, the contradiction between Michelson and the Aberration of Light disappears with this hypothesis, because, in the Michelson experiment, light would propagate in the ether atmosphere that accompanies the Earth in its movement, and in the case of light coming from the Stars, that light would propagate in the Interplanetary Ether. It is also worth remembering that Relativity, like all theories, may be wrong even though experiments confirm its consequences.

The fact that observations or experiments agree with the consequences of Relativity is not, therefore, a decisive argument in favor of this Theory. The Theory of Relativity is, therefore, subject to the possibility of being rejected, like so many other theories have been after they were considered definitive, but it would always have achieved solutions to problems that had previously been unsolved.[79]

Bibliography

Cabeza, C., et al. n.d. *About Einstein's Visit to Uruguay*. Vol. 6, 21–37. (without editorial data or date). Via: https://www.fing.edu.uy/biblioteca/revistas/247637-1.pdf.

Canales, J. 2020. *The Physicist and the Philosopher. Albert Einstein, Henri Bergson and the Debate That Changed Our Understanding of Time*. Barcelona: Arpa Editores.

Fouillée, A. 1891. Les origines de notre structure intellectuelle et cérébrale: I.—le kantisme. *Revue Philosophique de la France et de l'Étranger* 32:433–466.

Galles, C. D., and R. Rivarola. 2012. Notes for a history of physics in Rosario. In *Physics and Argentine Physicists. Stories for the Present*, ed. D. Hurtado, 157–187. Cordoba: National University of Cordoba-Argentine Physics Association.

Gangui, A., and E. L. Ortiz. 2020. The physicist Enrique Loedel Palumbo in the scientific corridor Montevideo-Buenos Aires-La Plata: 1920–1930. *Revista de Indias 80* (280): 815–846.

Gangui, A., and E. L. Ortiz. 2022. The Institute of Physics of La Plata in the first post-war period: The qualified transmission of sciences to youth. *Superior Council of Scientific Research Asclepio* 74 (2): 1–10.

Geille Castro, A. 1907. Study on the precision of the equinoxes. *Evolution* 2:435–439.

Geille Castro, A. 1910. Interesting application of an old theorem. *Evolution* 5:138.

Geille Castro, A. 1924a. The theory of relativity. *Annals of the University* 33:114.

Geille Castro, A. 1924b. Particularities of the engineering profession. *Journal of the Polytechnic Association of Uruguay* 189:1–3.

Hill, W. S. 1941. *Nuclear Combustion*. Rosario: Publications of the Faculty of Mathematics, Physico-Chemical and Natural applied to Industry.

Hill, W. S. 1948. *Physics Course of the Faculty of Humanities and Sciences*, 12. Archive Faculty of Humanities and Sciences Education. Walter S. Hill File.

Holton, G. 1973. *Thematic Origins of Scientific Thought: Kepler to Einstein*. Cambridge, MA: Harvard University Press.

Krauss, C. R., and L. Laino, eds. 2023. *Philosophers and Einstein's Relativity: The Early Philosophical Reception of the Relativistic Revolution*. Vol. 342. Berlin: Springer Nature.

Kuhn, T. S. 2015. *The Structure of Scientific Revolutions*. Mexico DF: Fondo de Cultura Económica.

Loedel, E. 1929. Gravitation. *LETRAS Magazine* 1:14–19.

Martínez, M. L. 2023. Beginnings of the history of science in Uruguay: first courses and cultivators. *Claves. Journal of History* 8 (15): 97–119.

[79] Williman (1923, 31–32).

Monteiro, M. 2022. *The Family of the First Uruguayan Physicist*. https://fisicamartin.blogspot.com/2022/06/la-familia-del-primer-fisico-uruguayo.html

Moraes, D. 2019. *Einstein in Uruguay. Chronicle of a Historic Trip*. Montevideo: Ediciones B.

Otero, M. H. 1996. *Of fallacies and non-fallacies in Vaz Ferreira*. In *Essays on Carlos Vaz Ferreira*, ed. M. Andreoli, 47–58. Montevideo: Faculty of Humanities and Sciences of Education Udelar.

Palumbo, E. L. 1922. Renaissance. *Monthly Student Magazine* 16.

Palumbo, E. L. 1926a. Optical and electrical constants of sugar cane. *Annalen der Physik* 384:533–549.

Palumbo, E. L. 1926b. The speed of light in a gravitational field. *Contribution to the study of physical and mathematical sciences* 4:455–460.

Palumbo, E. L. 1926c. The shape of the space-time surface of a gravitational field, which originates from a point-like mass. *Physikalische Zeitschrift* 27:645–648.

Palumbo, E. L. 1934. *Verses of a Physicist*. La Plata: Olivieri & Domínguez Graphic Workshops.

Palumbo, E. L. 1936. *Logic and Metaphysics. An Introduction to the Problem of Causality*. Buenos Aires: Editorial "C.L.E.S.".

Palumbo, E. L. 1940. *Bio-Bibliographic Card. Enrique Loedel Palumbo. Archive National University of La Plata*.

Palumbo, E. L. 1941. *Elementary Physics*. Buenos Aires: Angel Estrada & Cia. Publishers.

Palumbo, E. L. 1949. *Teaching of Physics*. Buenos Aires: Kapeluz Publishing.

Palumbo, E. L. 1955. A good still strange to general culture: the theory of relativity. *Engineering Journal* 563 (March): 73–74.

Palumbo, E. L., and S. De Luca. 1940. *Cosmography or Elements of Astronomy*. Buenos Aires: Angel Estrada & Cia. Publishers.

Pernot, F., and A. Geille Castro. 1911. Scientific investigations. *Evolution* 2:45–47.

Queijo Olano, J. A. 2023. *From the Laboratory: The Origins of Physics in Uruguay*. Montevideo: Biblioteca Plural, University Editions. (Funded by CSIC, Udelar).

Queijo Olano, J. A., and A. A. P. Videira. 2022a. The ethics of scientific practice: Enrique Gaviola and physics in Argentina between 1930 and 1956. *Knowledge. Journal of the History of Sciences and Humanities* 5 (12): 89–113.

Queijo Olano, J. A., and A. A. P. Videira. 2022b. Walter Scott Hill and Uruguayan Physics. *Physics in Perspective* 24 (1): 35–71.

Reichenbach, H. 1959. *The Rise of Scientific Philosophy*. University of California Press.

Reyes Thevenet, A. 1920a. *Cosmography and Its Teaching*. Montevideo: Peña Hnos.

Reyes Thevenet, A. 1920b. Astronomy in science. *Pegasus Monthly Magazine* 23:418–427.

Rosenkranz, Z. E., & Einstein, A. (2023). *The Travel Diaries of Albert Einstein: South America*, 1925.

Scheps Grandal, J. G. 2008. *17 Records: Faculty of Engineering of Montevideo (1936–1938) by Julio Vilamajó, Architect*. Montevideo: Faculty of Architecture, University of the Republic.

Vaz Ferreira, C. 1940. *Illegitimate mathematical transcendentalizations and correlated fallacies*. Buenos Aires: University of Buenos Aires.

Vaz Ferreira, C. 1972. Number in tribute to Dr. Carlos Vaz Ferreira on the centenary of his birth. *National Library Journal* 6.

Von Reichenbach, M. C., and L. R. Andrini. 2015. A new form of quantified energy: Presentation of the Loyarte-Loedel controversy. *Knowledge and Time* 1 (1): 168–188.

Williman, C. 1923. *Relativity, Presented to the Preparatory Students of the University*. Imp, Peña Hnos: Montevideo.

Chapter 5
The University and Science Around 1950

As mentioned in the previous chapter, the institutional development of physics in the Faculty of Engineering and Related Branches is associated with the name of Walter S. Hill. From his entry as Head of Physics Laboratory, to the transformation of the laboratory into an Institute in 1941, his vision of physics was the vision that developed the Institute and the Udelar. As is made clear in another work, this conception of physics is very tied to the function of the laboratory as a space that went far beyond mere teaching[1]. As an example, we can take the article he published in 1940 in the *Anais da Academia Brasileira de Ciências*, titled "Absorption of fluorescence light by the fluorescent liquid" (Hill 1940). The first works that could be developed in Uruguayan laboratories, with scarce instrumental and input capacities, aimed primarily at the pedagogical work in a very specific way: teaching from the replication of experiments. These first steps in the laboratory of the Faculty of Engineering, which brought together professors, assistants, and students in the challenge of verifying results obtained elsewhere, were the first form of connection with the international world of physics. These practices oriented towards the replication of experiments had a lot to do with the material possibilities at this initial stage, marked by the lack of resources but also with the expectation of scientific development in contact with some developments that were practiced in the world. Following the example taken from the referred article, the theme of the absorption of light in fluorescent environments coincides with the studies on cathode rays that at the time led the French chemist Jean Perrin to win the Nobel Prize. That is to say, even though the exercise was not of intellectual creation, using the laboratory for the experimental training of students was also a way to stay updated on the problems that arose and appeared in the international bibliography.

Twenty-five years have passed since that first proposal by Kayel to create a Physics Institute, in 1941 the Physics Institute was finally installed, with Walter

[1] Queijo Olano (2023).

J. A. Queijo Olano, E. Mozo Meneses, *Giants Upon Our Shoulders: A Philosophical History of Physics in Uruguay*, Studies in History and Philosophy of Science 65, https://doi.org/10.1007/978-3-031-97729-9_5

S. Hill as director. The institute joined old institutes and laboratories that had been built at the beginning of the century in the Faculty: the Institute of Material Testing, in 1912, the Electrotechnical Laboratory, the Machines Laboratory and the Analytical Chemistry Laboratory, in the same year. During the 30s, these Electrotechnical and Analytical Chemistry laboratories will transform into institutes. In the 40s, in addition to the aforementioned Physics Institute, the Mathematics and Statistics Institute was created, of fundamental importance in the scientific tradition of the Udelar.

These institutional transformations did not mean an immediate attention to the requirements of the development of a fundamental science within the Faculty of Engineering. In more pragmatic terms, and for the case of physics, it meant the possibility of building a regional and international academic and scientific network for the development of the discipline. This was the great task that Hill undertook during all his years in charge of the direction of the Institute. His task of international networking allowed a growth that the own budgetary limitations of the Udelar would not have allowed in any way, and in this growth of the Institute, his figure acquired a relative relevance in the region and some more distant countries. Proof of this is that, in 1940, he was awarded the title of Foreign Academic Correspondent of the Royal Spanish Academy of Sciences. Also in those years, a long exchange began with an important official of the Rockefeller Foundation, Harry M. Miller Jr. This was the central link of his academic career, and it became a true friendship.

5.1 Institute of Physics and Theoretical Physics

Understanding the processes of the internationalization of science from developing countries implies, in part, understanding the role that international organizations, like the Rockefeller Foundation, had in the development of science in the post-war third world. In the post-war world, the United States initiated an international policy on several fronts on the continent, among which the work of expanding scientific capacities in South America was one of the central ones

> [The Rockefeller Foundation] supplemented this first-[2] hand information by systematically supporting travel by trusted experts, who then supplied reports of their observations and assessments, and by systematically collecting information from experts whose travels brought them into the vicinity of Rockefeller officers. At the same time, they sustained close advisory relationships with selected scientific elites, whose regular professional activity gave them a broader view of their disciplines that could inform Rockefeller programs.

The objective of the officers of this foundation was to identify the scientific leadership of each country and thus establish personal relationships with academic support and legitimacy. Thus, we can presume that it was through the recommendation

[2] Barany (2019, 257).

of Félix Cernuschi to several physics professors from prestigious universities that allowed Walter S. Hill to approach the foundation. This had already initiated contacts in the country through Rafael Laguardia, the first mathematician to whom the Rockefeller Foundation awarded a scholarship in Latin America. In a letter from early 1941, Cernuschi -who at that time was occupying a position at the Physics Institute of the University of Tucumán (UNT) -recommended two things to Hill: first, that he write to the Guggenheim Foundation suggesting the possibility of obtaining a study scholarship to travel to the United States, and second, that he mention as scientific references the Mexican physicist Manuel Sandoval Vallarta and the Moss Institute of the Massachusetts Institute of Technology (MIT), where he was working. Cernuschi had worked alongside Vallarta, being a young student, on the project of a research network on cosmic rays. That's why Cernuschi recognizes him as a "very good friend of mine" and about the quality of this relationship he affirmed his recommendation for Walter S. Hill to work in "theoretical physics"[3].

The letter from Walter S. Hill to Sandoval Vallarta, written at the suggestion of Cernuschi, unambiguously states the clear purposes he had for the development of the Institute in an initial stage: to work in "modern theoretical physics and nuclear physics"and to have an experience of working in a top-level laboratory like those at MIT, to, above all, know "the organization of the laboratory":

> In my country, very little Theoretical and Experimental Physics is still done. But currently I have an opportunity to organize a Physics Institute based on the current Laboratory of the Faculty of Engineering.[4]

The way in which Walter S. Hill intended to settle in the United States was through the Guggenheim scholarship, but Sandoval Vallarta, in his response—sent 10 days later—indicated that the deadlines to apply for such funds had expired a few months earlier. The Mexican physicist encouraged, beyond the above, two real possibilities of joint work: on the one hand, he announced his visit to Montevideo in the month of September of that year, and, on the other, he was at his disposal for any link that could be established. So much so that, in a new letter, in May, Walter S. Hill invited Vallarta to give some lectures from the Physics chair of the Faculty of Engineering. Closing the year 1941, Hill's correspondence shows that his intentions to increase the network of academic and scientific institutional links continued to be his main concern. Taking advantage of the stay in the United States of his personal friend, the former dean of the FIA Luis Giorgi, he wrote to him to make the necessary arrangements to get in touch with Enrico Fermi. There, he insisted again with the same:

> I am interested not only in working alongside a man like Fermi, but also in observing the organization of research and laboratories, to transplant what can be in these payments (especially Nuclear Physics, I aspire to create a small study center in these latitudes, what do you think?).[5]

---

[3] Hill (1941d).

[4] Hill (1941a).

[5] Hill (1941c).

It was towards the end of 1941 that contacts with Harry M. Miller and the Rockefeller Foundation became more frequent. The record of the correspondence between them indicates that it would have started on 1st of October of that year, when Miller wrote to the dean of the Faculty of Engineering requesting to initiate the necessary steps to be able to grant Walter S. Hill a one-year scholarship to work at some prestigious university in the United States (a scholarship requested along with that of the mathematician Rafael Laguardia). The negotiations would have been previously agreed upon during Miller's visit to Montevideo at the beginning of that same year, and that note only had the pretensions of starting the administrative procedures. The agreement was that Walter S. Hill could settle in the United States and work alongside Sandoval Vallarta (at MIT) or Enrico Fermi (at Columbia University). A major obstacle presented itself, and it was the size of the Hill-Secco family (with five children). Even so, Walter S. Hill harbored optimistic hopes of being able to solve his situation if the scholarship was granted to him: "But for my part, I will try to overcome them, as this opportunity is important for my career and, above all, to promote and prestige the research of physics in my country"[6].

In January 1942, Miller wrote again to Hill, announcing that neither his scholarship nor Laguardia's had been accepted and that the reasons for the rejection would be explained in another letter that Miller addressed to Dean Vicente García. The report that officer Harry M. Miller raised to the Rockefeller Foundation said the following:

> Prof. Hill wishes to spend a year working on nuclear physics in the U.S., not, however, on the type of nuclear physics studied with the help of a cyclotron and has no hope of having such equipment at his disposal upon his return to Uruguay. He has expressed his desire to work with Prof. Vallarta at the Mass. Inst. of Tech, and the officers of NS [Nova Scotia] will explore the possibility of him being received there or at some other suitable institution in the U.S. [7]

The reason why his scholarship was denied had to do, as had been mentioned in the correspondence between Hill and Miller, with the numerous family of the Uruguayan physicist. On December 31, 1941, at a conference of administrators of the Rockefeller Foundation, it was resolved that no study scholarship would be granted to applicants from Latin America who were married or had family dependents in their country of origin. In the records of the Rockefeller Foundation it is mentioned that Walter S. Hill's application was treated as a priority, but that the result in this regard had not yet been announced [8].

In October 1941, the Uruguayan Chamber of Deputies had approved the creation of the Physics Institute, which would be a reality by the end of that year.

---

[6] Hill (1941b).

[7] Hill (1942b).

[8] Hill (1943b).

5.2 Latin American Integration in Cosmic Rays

When the FHyC was founded in 1945, Hill held the Chair of Physical-Mathematical Sciences that was created there. He came to this position in 1937, winning the competition that the IES had called for a professor to teach the courses of this subject. In the ruling that is part of the resolution of the Udelar Council, it is said that in Walter S. Hill not only do the merits and works stand out in relation to the other applicants in the competition, "...but also, they reveal in their author original talent as a researcher and exceptional conditions as a man of science" (Faculty of Humanities and Sciences, 1937). In the Chair of Physical-Mathematics of the FHyC, and not without certain difficulties, he continued his work in the construction of the international networks of physicists that he had already promoted in the Faculty of Engineering. Due to its organizational structure based on chairs, there were very few possibilities to form any laboratory and this was what, in the long run, ended up distancing Walter S. Hill from this institution. His vision of science held that the curricular structures of the chairs only make sense from the existence of the laboratory, and that is why all the chairs under his charge - in FHyC or in the Faculty of Engineering - structured their teaching with the laboratory as the axis. The symbolic and practical position of the laboratory thus structured all academic activity.

On August 24, 1942, he received a letter from the Italian-Russian physicist Gleb Vassielievich Wataghin[9] who was based in that Brazilian city and had developed a fruitful laboratory at the University of São Paulo (USP). Thanks to Miller's visit to that laboratory and the good recommendations he left for Walter S. Hill, Wataghin sent an invitation for the Uruguayan physicist to spend a research period at that center. The Rockefeller Foundation seemed to closely follow the steps of this laboratory led by Wataghin. Hill's response to the invitation arrived a few days later, on the one hand, accepting the possibility of visiting the USP, and, on the other hand, making explicit the difficulties that existed in Montevideo for the practice of physics. He himself, in that letter, recognized himself more as a theoretical physicist not by choice, but by resignation, but encouraged the possibility of developing a laboratory oriented to nuclear physics and X-rays.

> Our Institute is of recent creation (January of this year) and was founded on the basis of the Physics Laboratory of the Faculty of Engineering.- This laboratory is discreetly equipped in some aspects of Physics, for example, spectrography and X-rays.- But unfortunately I have almost no collaboration and also, the resources assigned in the budget, are very reduced.[10]

---

[9] Gleb Vassielievich Wataghin (1899–1986) was an Italo-Russian physicist who played a central role in the construction of physics in Brazil, starting from his participation in the University of São Paulo. His name was recommended by Enrico Fermi, with whom he worked, and his career shows us that he was formed in the fermenting European environment of the early decades of the twentieth century. A complete overview of his academic life can be found in Tavares, Bagdonas and Passos Videira (2020).

[10] Hill (1942a).

During the course of that year, he sent Wataghin the books he had published up to that point, and pointed out that he kept alive the hope of settling for a while in São Paulo with his whole family. One of those books was *Nuclear Combustion* . It is important to note that the issue of nuclear energy, the physical enigmas of nuclear fission and energy control, were the subject of unusual reserve in the international academic field. In the United States, a censorship was established from April 1940 on the publication of research on uranium fission; which also occurred in all the warring powers, in order to control the scope of the advances that were obtained. Neutral countries had greater freedom in this regard. That is why Hill's work was of relative relevance in the region. Hill pointed out there the importance of the discovery of Hahn and Strassmann, after which it was understood that it was possible to achieve the chain reaction, releasing a large amount of energy in a short space of time, which "makes the process have the characteristics of an explosion". Hill mentions the work of Francis Perrin in which the critical mass is estimated at about 12 tons of uranium. Undoubtedly, Hill's article can be considered pioneering in the dissemination in our environment of this subject.[11]

On November 18, 1942, Harry M. Miller confirmed that the Rockefeller Foundation would provide the budget for a scholarship stay in São Paulo for the following year, even though suspicions about Walter S. Hill's difficulties in fulfilling his stay were known in advance. And, indeed, Hill did not travel for two reasons, which he made explicit in a letter to Miller: the situation of the incipient Institute of Physics required his presence in times of few resources and need for growth, and, in addition, the works in the Faculty of Engineering, for which he was responsible along with the architect Julio Vilamajó, also required his presence (Hill, 1943a). In addition to these reasons, at the end of that year, the Executive Power of the government of Alfredo Baldomir appointed him to integrate a commission that would study aerodromes in the United States and Brazil with a view to building what would be the Carrasco International Airport (inaugurated in 1948). Hill would be the calculating engineer of that work and his project began with a tour of the most important American aerodromes.

It was not in 1943 that Walter S. Hill traveled to Wataghin's laboratory. However, despite desisting on that occasion from the scholarship, relations with Miller and the Rockefeller Foundation remained firm. At the beginning of 1943, he traveled to the United States on the official mission mentioned and Miller wrote to MIT so that he would be received by the American physicist John H. van Vleck[12]. In those years, there were The Manhattan Project had begun and the most important physicists working in the United States had been recruited for the purpose of developing

---

[11] Galles and Rivarola (2012, 160–161).

[12] John Hasbrouck van Vleck (1899–1980) was an American physicist who worked at the universities of Wisconsin and Harvard. He received the Nobel Prize in 1977 for his contributions to laser beam research. He was one of the many renowned physicists who were part of the Academic Committee of Los Alamos, as well as the previous meetings held in Chicago, aimed at the development of the atomic bomb in the USA.

weaponry from recent discoveries about nuclear fission. Van Vleck was one of those physicists summoned by Robert J. Oppenheimer to join the arms race that developed during that decade between the North American country, England, and Germany. Although that visit was not possible, the figure of Walter S. Hill was thus introduced into the world of American physics by Harry M. Miller:

> Dear Dr. Van Vleck
>
> I am writing to introduce you to Professor Walter S. Hill, Director of the Institute of Physics at the Faculty of Engineering of the University of Montevideo, Uruguay. Professor Hill is a theoretical physicist who would especially like to meet you and have the opportunity to talk with you and some of your colleagues during a visit he hopes to make to Boston in the next week or ten days [...].Professor Hill is a distinguished Uruguayan physicist who hopes to spend the next year working alongside Professor Wataghin in Sao Paulo, under the auspices of the Rockefeller Foundation. He is a man of broad interests and a charming personality and I am sure you will enjoy meeting him and discussing matters of mutual interest. Any attention you can give to Professor Hill and his friends will be greatly appreciated. [13]

The development of the Institute of Physics showed signs of good progress in terms of equipment, reaching the end of 1944[14]. On the one hand, they were waiting for the arrival of a Geiger-Müller counter donated by the American company Westinghouse Electric Company, which was of vital importance for the laboratory, as the university's budget situation continued to be problematic. Walter S. Hill mentioned to Miller that, with the arrival of this equipment, he wanted to develop research in cosmic rays. On the other hand, the Institute had managed to hire two full-time interns, who helped Hill with the tasks: a former student of his, Enrique de Martini (future dean of the Faculty of Engineering) and Laura Levi, a PhD in Physics in Italy and daughter of the prestigious Italian mathematician Beppo Levi. From 1941 to 1945, Hill seemed to have a clear idea that the line of research in cosmic rays could be developed in Montevideo. This confidence of Hill, over those years, seemed to be more the product of the enthusiasm of the new project than of a founded and substantiated appreciation of what cosmic ray research meant. It is also possible that the main scientific environment of the continent, involved in this particular type of research, also encouraged this confidence in the new director of the Institute of Physics. In this sense, it is worth observing the report of the year 1945 on the lines of research that occupied Hill reaching the middle of the century, this time from his chair of Physics-Mathematics in the FHyC.

> The reorganization of the radioactivity section was initiated, which has been permanently installed in a room of the Institute of Physics. — We currently have a Wulf electrometer, a Hoffmann elect., a Lindemann, an ionization chamber, an Engel fontanoscope, several common electrometers, a precision variable capacitor, a stabilized high voltage transformer, etc. — A Geiger Mueller counter is under construction. — Apart from purely speculative activities in this chapter, the Institute has organized the section of radioactivity measurements of native minerals.

---

[13] Miller (1943b).

[14] Hill (1944a).

> A cosmic ray laboratory has been annexed to this section with the purpose of adhering it to the International Institution for the study of Cosmic Rays. — For this reason, conversations have been initiated with foreign researchers.
>
> The interest of this initiative lies not only in introducing this kind of studies in our environment, but in the need to provide local data for a better understanding of the geographical and temporal distribution of cosmic rays. — The location of Montevideo (lat. 34951' South), is exceptionally suitable to form one more link in the chain of laboratories studying cosmic rays in South America.[15]

Although in his reports and correspondences Walter S. Hill seems to indicate his intention to develop research in cosmic rays, none of this seems to have occurred. Some reasons seem to explain this apparent contradiction. Firstly, we must be clear about what implied and why it was so relevant in those years the research in cosmic rays [16]. The study of these fields begins with the existence of certain "residues" of air ionization that could not be attributed to known phenomena, such as radiation. It is a problem that arises in the first decades of the twentieth century, to find the causes of these residues, and one of the first hypotheses rests on the existence of the influence of cosmic rays on our atmosphere. But this task presents a great challenge: the problem of precision and the conditions for the measurement of these rays. In Hill's 1945 report, the physicist recounts the geographical peculiarities of Uruguay for this type of research, because precisely one of the substantive changes that cosmic ray research offers is the need to study them in different latitudes and geographical areas that would allow a better understanding of their nature. In a classic text of modern physics, Nobel laureate Ernest Pascual Jordan allows us to understand the depths of what cosmic ray research involved:

> (...) in recent years, this study has become one of the most laborious fields of physical research. Observations and measurements have been accumulated through numerous sea voyages, through special expeditions that, from Spitzberg to New Zealand and Tierra del Fuego, have traversed all the oceans trying to determine the geographical distribution of the intensity of these cosmic rays on the surface of the Earth. The devices have been taken to the highest mountains, for example, the Alps and the Peruvian Andes, with the aim of seeing how these radiations behave at these heights (...) At these heights, cosmic rays, not notably weakened by the Earth's atmosphere, offer an intensity two hundred times greater than sea level. In other cases, measuring instruments are placed in the sea, hundreds of meters deep: the weakest traces of cosmic radiation were recorded, corresponding to a very special portion of these rays endowed with special penetrating power.[17]

The expeditionary nature made cosmic ray research a highly collective and international project, and that is why Hill's interest in getting involved in it is explained. His academic networks, from Vallarta to Wataghin, influenced by the direct

---

[15] Hill (1946, 17).

[16] We appreciate the invaluable contribution of Professor Adriana Minor, from the Center for Historical Studies of El Colegio de México, to clarify in depth the history of the development of cosmic rays in Latin America.

[17] Jordan (1953 [1938], 149–150).

influence of the Rockefeller Foundation, drove him in this direction. In general terms, Latin America played an extremely relevant role in these investigations from the United States, which covers more dimensions than scientific ones, such as -for example- the geopolitical dimension.[18] This impulse found its highest point of expression in the discovery by the Brazilian physicist César Lattes of the π-meson, whom Hill himself will nominate as a candidate for the Nobel Prize along with the American physicist Eugéne Gardner, both responsible for the discovery in 1949.[19] But on the other hand, as already pointed out, the practice in physics that most interested Hill was the one that could be developed within the laboratory, which in a certain way made him incompatible with the "field" investigations required by the study of cosmic rays.

As the institute grew —slowly—, Walter S. Hill's academic reputation also grew, and research in cosmic rays provided a foothold to expand Uruguay's academic and scientific networks, despite the fact that no concrete results are recorded in this field. This integration and recognition of the Physics Institute, and of Hill as its director, can be glimpsed in the reference role he played in front of the Rockefeller Foundation. In 1944, when writing a letter to his friend Harry Miller about the procedures concerning the acquisition of equipment for the institute, he allowed himself to make the following suggestion:

> I now wish to speak to you about a young mathematician who is interested in a study grant.
>
> This is Professor Eng. José L. Massera, 28 years old, graduated from the Faculty in February 1943, who is practically in charge of the Mathematics Institute, which is directed by our friend Laguardia, while he is abroad.
>
> He is a young man of exceptional talent and culture, especially in mathematics. He carried out brilliant studies at the Faculty and subsequently showed a great vocation for teaching. Since 1937 he has been a professor of Mathematical Analysis I.
>
> Knowing the merits of Prof. Massera and knowing that his current aspiration is to expand his knowledge abroad, I thought it was convincing that he established a link with the Rockefeller Foundation.- For this reason, he will write a letter to you, expressing his wishes. I trust that you will consider his request.[20]

This recommendation of the mathematician José Luis Massera[21] was added to the one that the previous summer had been made by the Rockefeller official, Marshall Stone. Without a doubt, Walter S. Hill was, at that point, a trusted reference for the foundation.

[18] Minor (2020).

[19] Nascimento (2015).

[20] Miller (1944b).

[21] José Luis Massera (1915–2002) was one of the main mathematicians not only in Uruguay, but probably also in Latin America. Among his most important contributions to the study of equilibria in differential equations, the theorem that establishes the reciprocal to one of Lyapunov's theorems stands out. But, in addition, Massera was a tireless fighter for the formation of a community mathematics in Uruguay, as well as a political reference in the Communist Party of Uruguay.

In 1944, Walter S. Hill's academic networks continued to expand: at that time he came into contact with the Austrian physicist Guido Beck[22] who by then resided in Córdoba and worked at the National Astronomical Observatory of that city. The link with Beck involved, in the first place, the possibility that Hill could join the meetings that the Association of Physicists of Argentina (AFA) held in La Plata. Despite showing an initial interest in participating in them, Hill finally did not manage to go, without thereby ceasing to continue his correspondence with Beck. On October 17 of that year, Walter S. Hill sent him a letter informing him that he had learned about Beck's desire to move to Montevideo to work, so he immediately made the institute he directed available to him, as well as his commitment to start the procedures for Beck to be able to settle in the Uruguayan capital. On September 13, 1945, the Austrian physicist was invited by the Physics Institute of the Faculty of Engineering and gave a lecture "The electromagnetic analogue of the sound of a projectile (Cherenkov effect)".

Towards the end of 1945, a new letter from Miller encouraged Hill to visit the most prestigious physics laboratories in the United States. That year the support that the Rockefeller Foundation would be allocated to the laboratory of the Institute of Physics, which was divided into an advance of 1500 pesos that the faculty put towards the purchase of materials and 1600 dollars that Walter S. Hill requested to be deposited in an American bank so that they could be used for the purchase of equipment in that country.

The contacts with Wataghin and the mediation of the Rockefeller Foundation finally allowed, in February 1947, Walter S. Hill to conduct a research stay at the USP for a month. The visit had such an impact that it would be mentioned in the first letter to his friend Miller once he returned to his activities in Montevideo. From this letter, some elements emerge that help to understand the changes that Hill would introduce at the Institute between 1947 and 1948. In addition to working in a top-notch physics laboratory, perhaps the most important in South America at that time, he also regularly interacted with a group of renowned physicists, such as Marcelo Damy de Sousa Santos, Mário Schenberg, and Hans Stammreich. This contact with some of the most significant physics in the region deeply impressed Hill, who —as far as we know— had been able to develop physics in Uruguay based on a few contacts by correspondence and the solitude of someone trying to develop the discipline within a faculty so focused on the training of professionals as was the Faculty of Engineering.

This stay meant the possibility of defining with greater certainty those directions that should be established for the Institute of Physics. A first substantial aspect seems to have been the need to mark thematic lines for the laboratory. The dispersion of interests seemed to be a characteristic element in the *scientific personality* of

[22] Guido Beck (1903–1988) was an Austrian physicist who, from the second half of the thirties, settled in Argentina and became an important actor for the development of theoretical physics in that country, through joint work with Enrique Gaviola at the National Astronomical Observatory of Cordoba. In the fifties, he moved to Rio de Janeiro to join the group of physicists working at the Brazilian Center for Physical Research.

Walter S. Hill. Until then, he had published on matters of wave and particle physics, fluid mechanics, general equilibrium equations, general thermodynamics, flat concrete slabs with crossed reinforcements, studies on the properties of fluorescence light. This thematic breadth, this dispersion of interests, changed after his stay in São Paulo since, at least declaratively, Hill announced a redirection of research towards a few topics. In fact, this was one of the elements that he most valued from his work with Wataghin.

> I am sure that in a very short time, the contribution of the Physics Department will be very important and select. Dr. Wataghin was very fortunate in his idea of concentrating efforts in well-defined fields with great possibilities, such as cosmic rays and Nuclear Physics.[23]

The visit to the laboratory in São Paulo was, then, in addition to a space for personal training and thematic definitions for his own route in physics, the space where to recognize the direction that quality scientific activity should take to achieve relevant objectives.

A year later, the First Conference of Latin American Scientific Experts was convened by UNESCO, to be held in Montevideo. The newspaper *El Bien Público* dedicated a few paragraphs to the conference on the same day of its inauguration. It mentioned the presence of the outstanding researchers who arrived in Uruguayan lands from Brazil, Colombia, Ecuador, El Salvador, the Dominican Republic, and Venezuela, in addition to the Uruguayan delegation, made up of Walter S. Hill, Clemente Estable[24]—who would preside over the meeting—, Rodolfo Tálice and Germán Villar. The conference began on Tuesday, September 7, 1948, in the Paraninfo of the University and would end on Friday of that same week.

A few days before the conference, a group of Uruguayan scientists —among whom Félix Cernuschi, Segismundo Gerszonowicz, Rafael Laguardia, José Luis Massera, Rodolfo Tálice, Francisco Sáez, Raúl Vaz Ferreira, and Walter S. Hill stood out— published an article in the magazine *Ciencia e Investigación*, of the Argentine Association for the Advancement of Science. The article was titled "Suggestions for the meeting of Latin American scientists to be held this month in Montevideo" and presented a series of recommendations and actions to be carried out in the immediate future once the conference was over. A first request was aimed at establishing a Latin American Association of Scientists, made up of people "animated by a fervent desire to promote a broad development of science in our countries, and organized in such a way that guarantees the efficiency of its operation and prevents it from becoming a kind of inert honorary academy".[25]

---

[23] Hill (1947).

[24] Clemente Estable (1894–1976) was a pioneer in Uruguay in the field of biological research. From a scholarship obtained in the twenties, he managed to travel to Madrid to study at the Institute of Histological Research, under the direction of Santiago Ramón y Cajal. Through his position as director of the Biological Sciences Laboratory of the Primary Education Council he made important scientific contributions. A more complete biography and documentation can be seen in the file of Clemente Estable from AGU (<http://historiasuniversitarias.edu.uy/biografia/estable-clemente/>).

[25] Bonnecarrére (1948, p. 377).

In the note, the signatories indicated some actions that could be taken (an academic journal *of scientific information*, for example), as well as the problems facing the community reaching the middle of the century. On the one hand, problems related to the teaching of science and its public understanding; on the other hand, problems regarding the role of science in the progress of Latin American societies. To solve the issues derived from the problems of teaching science and its public understanding, the document suggested conducting a survey among scientific institutions and experts that would allow obtaining a first picture of the situation of science in Latin American countries. Regarding the solutions to the problems that arise from the development and progress of science, the measures suggested ranged from strengthening the scientific libraries of universities to providing scientists with budgetary capabilities so that they could achieve exclusive dedication. It is interesting to note, for the Uruguayan case, that this concern for the almost non-existent recognition of Science in Montevidean society, still rooted in the intellectual weight of the humanities and letters, will be a constant axis of several scientists in their public participation, perhaps the most paradigmatic case being that of Félix Cernuschi.

Hill's participation in such an event is known through a small sheet found among his documents, where there are some brief notes prepared for the conference, presented with the title "Motion of Walter S. Hill". Basically, his contribution was in a very specific aspect: to support the idea of carrying out a survey of each of the "fissionable materials"in Latin America, which would be "very important for the future of nations, as these investigations are related to the obtaining of energy sources, and for the progress of pure science, as they involve studies of the highest interest for knowledge".[26]

As international advances in the world of physics became a reality, the search for radioactive minerals became part of the state policies of many countries. Latin America was not alien to this race, especially Brazil. International interests in the detection of minerals and the control of their trade in the world became matters of international relevance. Walter S. Hill's motion seems to indicate something in this sense, although the available documentation allows no more than these assumptions.[27]

Two months after the conference, in December 1948, the Uruguayan Association for the Progress of Science (AUPCC) was founded, following the development model promoted by UNESCO in all third world countries (Abarzúa, 2017). The Uruguayan academic community fully involved itself in the initiative and was an active participant in the Association, which, like in other places, proposed three categories of affiliation: corresponding member, titular member and adherent member. Walter S. Hill was one of the signatories of the founding act as a titular member, although his participation in the daily functioning of the association seems to have been little. The first issue of the association's bulletin, edited in 1952, includes a review co-authored by him, along with Dr. Estenio Hormaeche, who was the director of the Institute of Experimental Hygiene of the Faculty of Medicine. Together

[26] Hill (1948).

[27] Ribeiro de Andrade and Muniz (2006).

they were the delegation that the AUPCC designated to be represented at the III Annual Meeting of the Brazilian Society for the Progress of Science, in whose report they highlighted two central aspects of this meeting organized in Brazil: on the one hand, they highlighted a general discussion that was of interest to all those present, related to the possibility of applying psychotechnical tests for the selection of technicians and scientists. On the other hand, they were surprised by the plurality of disciplines integrated in the meeting, which in their opinion was more of a scientific congress than a meeting between Latin American scientific institutions:

> In reality, more than an annual meeting of the SBPC, it was a scientific congress organized by the same, and we say this without any spirit of criticism since in the final analysis the realization of a scientific congress is always a means of stimulating production. We also believe that this meeting has been different from the usual ones precisely because by encompassing so many different sciences it has had the virtue of bringing into contact researchers engaged in the most diverse disciplines, with indisputable advantage for all.[28]

In summary, by the early 50s, Walter S. Hill had broad regional and international recognition in the community of physicists. In the FHyC, his Chair of Physical-Mathematical Sciences showed a movement in its academic activities that, in a way, accounts for the networks that Walter S. Hill created: during that period, at the request of Hill, Dr. Néstor B. Cacciapuoti and his assistant, engineer Irene Weinkrantz Weinstock de Ewenson were invited by the dean Carlos Vaz Ferreira.[29]

5.3 From Nuclear Physics to Nuclear Medicine

The Conference of Scientific Experts organized by UNESCO was an event with profound consequences for Walter S. Hill. There he established new academic relationships that would mark his work in the laboratory towards new directions. This new thematic perspective was strictly in tune with the role and direction that UNESCO suggested for the development of nuclear physics. Of course, these changes were not simple coincidences between Hill's interests and those of UNESCO. The use of nuclear energy for peaceful purposes was a motive for

---

[28] Hormaeche and Hill (1952, 16).

[29] Irene Weinkrantz Weinstock de Ewenson (1917–2007) was a Jewish Polish physicist who managed to survive the Nazi persecution during World War II, escaping from the Warsaw ghetto she moved to Uruguay. Having studied physics before 1938, she began collaborating at the FIA institute and was the one who replaced Walter S. Hill in his direction when he retired. Irene de Ewenson, as she is referenced in the sources worked, was a student of Hill both at the FIA and at the FHyC, and participated in the work at the Scientific Research Laboratory of ANCAP, in 1949 (Hill 1949), and taught courses on theoretical and experimental spectrography. Her story, still lacking documentation, deserves a deeper investigation for two major reasons: in the scientific environment in general, and that of physics in particular, the presence of women was very scarce; on the other hand, the Polish origins of Irene Weinkrantz coincide with a Polish generation from the beginning of the twentieth century trained in analytical philosophy, mainly composed of physicists and mathematicians (Sznajder 2024).

incentive and propaganda from this international organization for the development of applied nuclear research in several countries. In particular, its medical and biological use was a deliberate political strategy to show the positive consequences of nuclear research beyond its warlike applications.

> The early postwar publicity about the medical breakthroughs that radioisotopes would bring focused on their therapeutic uses, particularly in the treatment of cancer. The hope that isotopes would cure cancer was predicated on the notion that they would localize to specific tumors and deliver internal radiation. However, the expectation that radioisotopes would become so-called magic bullets to fight cancer didn't fully materialize, whereas radioisotopes did become important tools in the area of medical diagnostics.[30]

On June 12, 1951, Walter S. Hill wrote a letter to Dean Carlos Berta, in which he notifies him about the plans he envisioned for the Physics Institute.

> The work initiated in 1949 [...] related to the radioactivity of some minerals in the country, gradually led us to a broader study, encompassing other physical proposals of our minerals such as refractometry, electrical and magnetic susceptibilities, structural determinations, etc.
>
> [...] In particular, I consider it necessary to intensify everything related to the structural research laboratory, by the methods of X-ray and electron diffraction.[31]

In the same letter, Hill indicated to the dean the path he thought to take to make these research lines viable, and this strategy showed UNESCO at the center. He related there that, knowing the budgetary difficulties that Udelar was experiencing, at the conference he had initiated contacts with UNESCO leaders to undertake "systematic studies" on the subject and "complete our laboratory in terms of scientific equipment and personnel". This last point was of vital relevance, because it involved a whole new series of contacts with European laboratories, as well as the first clear and firm experience of training genuine human resources of the Institute: the case of the young Stephenson Caticha Ellis. But, on the other hand, it did not seem so clear in those years the panorama that the FHyC offered him. Undoubtedly, the educational project that was being built there differed in substantial terms with the way in which Walter S. Hill conceived what should be the scientific training of students. The facilities that the Faculty of Engineering guaranteed him, where —as has been said— the laboratory occupied the center of its pedagogical proposal, did not coincide conceptually with the novel educational project that was beginning in Humanities and Sciences, more oriented towards a professorial and encyclopedic training. In 1951 the lawyer Justino Jiménez de Aréchaga assumed the deanship in that Faculty and in March of that year Hill wrote him a long letter in which he expressed his doubts about the course of Physics that he taught in that house of studies. As few times in his notes, we can see reflected in this letter some personal notion about what it means *to do physics* and, for that reason, to stop at some of its passages is worth it since it is one of the first proposals for structuring physics studies programmatically, considering the necessary conditions for such project.

[30] Creager (2014, 45).

[31] Hill (1951b).

> Physics is an extensive and complex science, which covers many topics, which constitute as many specialties united by the nexus of a common scientific method. [...] To fix ideas, I will indicate one of the possible divisions into basic subjects:General Physics, Rational Mechanics, Analytical and Relative Mechanics, Quantum and Statistical Mechanics, Atomic Physics and Nuclear Physics, Electronics, Electromagnetism, Thermodynamics and Kinetic Theory.
>
> This set of subjects or signatures could constitute the foundation of a bachelor's or even a doctorate in Physics, complemented with other cultural formation subjects, such as Philosophy, Mathematics, History of Sciences, etc.
>
> Naturally, a body of specialized professors would be required, laboratories and sufficient economic means to carry out research work, without which, teaching is fatally affected.
>
> [...] In reality, none of my courses intended or could be part of the core of courses of a bachelor's or a doctorate.
>
> They were isolated courses for a student body of unknown previous training, which necessarily had to adjust to their average capacity, and also to circumstances of the moment.
>
> I consider that this situation cannot and should not continue. I consider necessary to provide the Faculty of Humanities and Sciences with a body of professors and regular means to carry out a permanent teaching, susceptible to be organized in the usual university terms, whether a bachelor's or a doctorate.[32]

The concrete proposal was to start courses that, as they were completed, could be credited to a future Bachelor's degree in Physics. The interesting thing is to show how, as this young faculty consolidated, physics training could no longer be sustained either as a set of subjects useful for engineering or as a general course in a FHyC. The dizzying process of specialization that physics was experiencing in the first half of the twentieth century forced its teaching to become independent from the training of engineer to establish itself autonomously as a discipline differentiated from the rest of the natural sciences.

Hill's work in the study of the structures of minerals from Uruguayan soils continued its course in the following years. In January of 1950 he published an article in the magazine *Science and Research*, of the Argentine Association for the Progress of Science, with the title "Radioactive elements of fossil bones from the Tertiary and Quaternary"[33]. It is a preliminary study on the existence of radioactive elements in fossil remains of the region, in which he tests some hypotheses about the why of the high radioactive charge detected in the fossils of the mentioned periods. The work shows the experiences in which the Institute of Physics was beginning, which were moving away from those projections more linked to the study of theoretical physics or the study of cosmic rays.

Similarly, in 1953 appeared in the *Bulletin of the AUPCC* A report written by Hill about the early trials that the Physics Institute was conducting with the *Scintiscanner*, an automatic radioactivity recording device, built at the Institute itself. The report accounted for the techniques of radioisotope detection, already applied in the

---

[32] Hill (1951a).

[33] Hill (1950).

medical field, specifically in the detection and characterization of the thyroid gland. The report also announced trials on brain tumors and even the potential of the device for the study of matter by gamma radiation.[34]

To this academic reality should be added what was beginning in Walter S. Hill's personal life around the same years. In parallel to his academic development, he led a professional life that does not seem to have been minor. We know his work as an engineer, more precisely as a calculator in large works. We have talked about his participation in the construction of the new building of the Faculty of Engineering, his participation as an engineer in the works of the Carrasco International Airport and we also know about his work on the dome of a gymnasium of the Olimpia Club of Montevideo (along with the Uruguayan architect Eladio Dieste). However, from 1950 he began to develop a new venture: a private isotope laboratory. We know about this matter thanks to the published memoirs of a Romanian who arrived in Montevideo in 1940, exiled by the war, and who through a contact began to work with Walter S. Hill. This Romanian, an engineer registered upon arrival in Uruguay under the name of Enrique Coman, had studied at the University of Caen, in France, and the then rector of that university, Jean Valés, personally knew the Uruguayan professor Clemente Estable. When Coman arrived in Uruguay with a letter from the French rector addressed to Professor Estable, he immediately suggested that he approach Hill.

Both began a business partnership together. Enrique Coman had the international contact to import isotopes, while Hill contributed the scientific contribution. Both believed that the activity had a promising future not only in the scientific field, but also from a business perspective, because no one in the country was working with this technology in the medical field. They founded the laboratory, naming it with Einstein's emblematic formula: MC2. The MC2 laboratory hired a doctor, an exiled Austrian named Kurt Lassus, and Walter S. Hill took care of the construction of the laboratory equipment. They sold their services to doctors and hospitals, and many of the country's most important doctors learned about the new radioisotope technology in this laboratory. Among them were doctors José Manuel Cerviño and Jorge Traibel.

> The first medical applications of radioisotopes in the country were carried out by Dr. Jorge Traibel, Medical Director of the MC2 Laboratory, through thyroid studies with I-131. Also Dr. Félix Leborgne (p) carried out at the same time studies of placenta and globular survival (Uruguayan Society of Biology and Nuclear Medicine, 2014).

This laboratory was, in the end, a problem for Walter S. Hill. On the one hand, the fruitful economic business they believed it would be, as Coman claimed, never was. The type of studies provided by the laboratory was very expensive for the budget managed by the health system at that time in Uruguay. In addition, on the other hand, in 1956, when Hill was to be reelected to his position, a problem arose in the

[34] Hill (1953d).

Council of the Faculty of Engineering and Surveying (FIA)[35]: a complaint arose about a conflict of interest between the work that Walter S. Hill was carrying out in the laboratory of the Physics Institute and the practice in his private laboratory, where they also conducted studies with radioisotopes. The provision, control and distribution of radioisotopes for scientific use had fallen under the orbit of the FIA, particularly the Physics Institute. In other words, the national activity on this technique was controlled by the Institute and, in this sense, Walter S. Hill's private laboratory, which in turn provided services to the faculty, constituted a conflict of interest. Beyond what seemed to be a clear situation of abuse of functions, the general climate of the FIA Council was to accompany Hill's reelection in office. Thus, with only one vote against, he was renewed in 1956 in his position as director of the Physics Institute[36].

The new thematic turns of the Physics Institute, in connection with UNESCO, became noticeable in the session of the Faculty of Engineering Council on May 6, 1952, when Walter S. Hill reported on his passage through this organization and how he had managed to obtain a significant amount of money for equipment and human resources. In a session that took place in Paris, Hill had been summoned to present his research project. The only requirement that UNESCO demanded to study any application was that "the project involved - directly or indirectly - is related to the economic development of the country", as the proposals were headed by the States themselves.[37] Walter S. Hill's presentation seems to have been successful enough for him to be notified of the acceptance of the proposal in the following days. This, as we have said, referred to the new direction of the research lines of the Physics Institute, related to the study of the crystalline structures of the country's minerals and soils.

The funding obtained included an X-ray diffraction equipment with a recording device of approximately 10000 dollars; a 2000 dollar precision goniometer; the arrival of a foreign specialist for 1 or 2 years (the time that UNESCO used to hire specialists of this type for work stays), and two scholarships to train Uruguayans in laboratories in other countries.

As the purchase of equipment had to be advised by the specialist, Walter S. Hill first dedicated himself to the search for this by touring different laboratories in Europe. There were many scientists and tertiary education institutions he visited:

---

[35] In 1954, the Faculty of Engineering and Related Branches changed its name to the Faculty of Engineering and Surveying.

[36] The almost unanimous support is not minor for the type of problem. The weight that students began to acquire in the The general university and the Faculty of Engineering in particular were growing as the sixties approached, and having student support speaks to the recognition that Hill had among this group. Even the students, who represented the most confrontational position among the orders that governed the FIA, resolved in assembly the following: "We express our absolute confidence in Eng. Walter S. Hill against the comments that have been made regarding the relationship of a private laboratory with the Institute of Physics" (Faculty of Engineering 1956).

[37] Faculty of Engineering (1952).

the polytechnic of Zurich, where physicist Paul Scherrer worked[38]; the Free University of Brussels; the Mineralogy Laboratory of the Faculty of Sciences in Paris, where he contacted Professor Jean Wyart[39]; the University of Lisbon, where he contacted Professor Julio Palacios[40], and finally, the University of Oxford, where he found the Nobel Physics Laureate William Lawrence Bragg[41]. That is, the moment he received the funding he had already started the contacts that would allow him, once again, to place himself in the specialized field of physics dedicated to crystallography. Despite this, the hiring of the international expert who would work at the Institute of Physics took much longer than could be expected, due to Walter S. Hill's reluctance to bring the right person to the institute.

In August 1952 Hill inaugurated a course with the Argentine physicist Ernesto Galloni[42] on X-ray diffraction, with funding obtained from the Faculty of Engineering that allowed the transfer of Galloni once a week until the end of that year. To inaugurate the course, Walter S. Hill wrote a few words[43], which remind us of the attempts of the Institute of Physics, since its arrival, to form a solid core of studies based on X-ray diffraction. On the one hand, in those writings he highlighted the constant idea of what it meant to him to train in physics: to accompany theoretical development with practice in the laboratory. On the other hand, he emphasized the sustained work he did in favor of the training of human resources in the area of diffraction, despite the constant difficulties that existed in the laboratory.

In a letter from the end of August, addressed to the UNESCO representative Zoel González Ruiz, he commented on some advances, among which was the course that Ernesto Galloni was teaching with acceptable success. Apparently, many of Hill's collaborators were working with the Argentine physicist, who felt very comfortable

---

[38] Paul Scherrer (1890–1969) was a Swiss physicist who worked at the Federal Polytechnic School of Zurich, after having trained in Königsberg and Göttingen with Peter Debye, Nobel laureate in Chemistry. From 1954, and after having chaired the Swiss Commission for Atomic Energy Studies in the previous decade, he was called for the implementation of the European Organization for Nuclear Research (CERN, for its acronym in English). Today there is an institute in that country that bears his name (<https://www.psi.ch/en>).

[39] Jean Wyart (1902–1992) was a French crystallographer, member of the Academy of Sciences of France and professor at the University Paris VI. At the time of Hill's visit, he was directing the mineralogy laboratory of La Sorbonne. Years later, he was president of the International Union of Crystallography.

[40] Julio Palacios (1891–1970) was a Spanish physicist who specialized in X-ray diffraction in the early decades of the twentieth century. In 1932, when the National Institute of Physics and Chemistry was created, Palacios was appointed head of the X-Ray Diffraction Laboratory.

[41] William Lawrence Bragg (1890–1971) received the Nobel Prize in Physics in 1915—along with his father, William Henry— for their contributions to the field of X-ray diffraction crystallography. After World War II, he worked as a physicist at the University of Cambridge, where he was a professor of experimental physics.

[42] Ernesto Galloni was an Argentine engineer. His work stood out, above all, in the field of crystallography. He was a researcher and president, between 1955 and 1958, of the National Atomic Energy Commission of Argentina and, like Walter S. Hill, was highlighted as a correspondent of the Royal Academy of Sciences of Madrid.

[43] Hill (1952a).

because —according to his words—"our current team is much superior to that of Bs. As."[44]. The other indication of progress was the relationships he had established on his visit to Paris: a new course began in September of that year at the Institute of Physics, led by Professor Maurice Françon[45], director of the Institute of Optics at the University of Paris.

However, the situation regarding the contract of the specialist who would establish himself in Uruguay to train resources in the area of crystallography remained a problem. Faced with this circumstance, the response that came to Hill from UNESCO was of great benefit for the Institute: in addition to two scholarships for students and 20000 dollars to buy the X-ray diffraction equipment, a new specialist was approved who would join that person who would move to work in Montevideo. Even so, Walter S. Hill's decision was delayed, while also —for bureaucratic reasons— the agreement between the Ministry of Public Instruction of Uruguay and UNESCO was not signed. By the end of 1952, these delays became heavier for Hill because the budget forecasts of the organization for 1953 changed radically: the United States —responsible for 60% of UNESCO's budget— announced the reduction of their contributions, which greatly conditioned the possibilities of those projects. This news moved him to use his immediate family contacts to unblock the situation that prevented the signing of the agreement. Thus, in January 1953 he wrote to his father-in-law's half-brother, Óscar Secco Ellauri[46], who at that time was working as a UNESCO official and had been, just a few years before, Minister of Education and Culture in the government of Luis Batlle Berres. The next day, he also wrote a letter to the ambassador of Uruguay in Paris, requesting his intervention in the matter, with the hope that he could intercede to end the formal procedures that prevented the disbursement of the money obtained. Two days later, he sent letters to engineer Luis Giannattasio and Professor José Luis Carbajal Victorica, close contacts to the Minister of Public Instruction. Also, on that date he wrote an extensive letter to the rector of the Udelar, Leopoldo Agorio, informing him of the situation and the academic relevance of the project.

> The project consists of organizing a laboratory for the study of CRYSTAL STRUCTURE.
>
> This laboratory would be located at the Institute of Physics of the aforementioned Faculty [of Engineering].
>
> The purpose pursued with our project would be the following:
>
> 1) To carry out a systematic study of the structural characteristics of the country's minerals, using X-ray diffraction and
>
> 2) To create a research center for crystal structure, with a view to scientific progress and technical advice on the subject.

[44] Hill (1952b).

[45] Maurice Françon (1913–1996) was a French physicist who mainly worked in the field of optics. He obtained a position as a lecturer, in 1938, at the Faculty of Sciences of the University of Paris.

[46] Hill (1953a).

> We understand that this project has a solid connection with the country's economic development. Firstly, these studies are at the base of the best and most complete knowledge of the minerals and raw materials of technological interest existing in the national territory, and will effectively contribute to the relaxation of the wealth of our subsoil. Secondly, the realization of the project, although it is fundamentally a scientific enterprise, it cannot be denied that it will serve to stimulate technical production, either by promoting the training of experts or through direct advice in the various specialties linked to these studies—all of this, with a favorable impact on the industry and the national economy.[47]

It is undeniable that, by 1953, Walter S. Hill's interests had greatly expanded and diversified, and from the fixed idea of developing theoretical and nuclear physics, there was a shift towards specific work on identifying crystal structures. Looking at a letter he sent to Miller that same year, we see that he lists the possibilities that the Crystal Structure Laboratory seemed to be opening up for him:

> So that you can appreciate it and fulfilling what was promised quite some time ago, I list below the works to which we are dedicating our effort.-1- Crystal structure of minerals and organic compounds by X-ray and electron diffraction. [...] My plan is to organize a center for structural studies based in Montevideo, oriented not only to mineral structure, but also to organic. These studies are complemented by others, such as magnetic and electrical susceptibility, etc.
>
> 2- Collaboration with the Faculty of Medicine, Endocrinological and Radiological Institute and with the Ministry of Public Health in everything related to the use of radioactive isotopes. [...] We are particularly interested in the localization of brain tumors by radio-di-iodo-fluorescence or other radioisotopes[48].

The Institute of Physics, with which the Rockefeller Foundation had already had development ties, was beginning a new stage of research in radioisotopes, especially in the area of nuclear medicine. If the research in the use of isotopes had derived, in the case of Walter S. Hill, towards a practice oriented to nuclear medicine and clinical physics, it had not been due to the merit and initiative of its director. The research lines of the institute, at that point, were much more influenced by the possibility of funding initiatives.

In mid-June 1954, Hill continued to move his influences to achieve the approval and signature of the project by the national authorities, with a view to it being approved by the UNESCO Technical Assistance Board in July. The conditions had not changed: the project granted in a first stage the necessary equipment to install a crystallography laboratory (destined to the study of the minerals of the Uruguayan soil) and the visit of an expert for a period of 1 year. The counterpart of the Faculty of Engineering was to cover the salary and expenses of said expert, as well as to maintain the facilities and staff of the Institute of Physics. The second stage of the agreement committed UNESCO to send two members of the institute's work team to train abroad. This second stage would begin to be implemented towards the end of that year, when the UNESCO Technical Assistance Board asked the university in

---

[47] Hill (1953b).

[48] Hill (1953c).

general and the Faculty of Engineering in particular to think of names that could take advantage of the scholarships that UNESCO offered for training. In November, Walter S. Hill wrote a letter to the Minister of Foreign Affairs of Uruguay, together with the president of the UNESCO Technical Assistance Board, Dr. Miguel Albornoz, insisting on the need to advance in the signing of the necessary documentation so that the project could finally be realized in 1955.

5.4 New Directions: Crystallography

Thus ended the year 1954. On the one hand, in the persistent struggle to advance, given the bureaucratic obstacles that represented getting the UNESCO-led crystallography project underway. In this sense, the first step that was achieved was to propose sending engineer Caticha Ellis to train abroad. On the other hand, aiming at the possibilities that the Rockefeller Foundation seemed to offer him once again, Hill had to assume the commitment to advance in studies that linked physics and biology, more precisely in the research of radioactive isotopes. The other significant fact at that date was that the whirlwind of work that these two funding sources and the follow-up of their plausible realization made his performance as a teacher in the FHyC subject to criticism.

In a note of September 1st, Dean Vaz Ferreira transcribed the reasons that Walter S. Hill had given to the Council to present his resignation, which had been accepted by that body. The note, addressed to the dean, said:

> I am pleased to inform you that the work I have been carrying out with my collaborators, related to radioisotopes, have acquired such volume at this moment, that they practically absorb all my time. This situation has also been complicated by the increase we are giving to our X-ray diffraction laboratory, due to its possible transformation into a Center for Crystal Structure Studies through X-ray diffraction. with the cooperation of UNESCO. For such reasons, I am forced to restrict my direct intervention in regular courses to the maximum. Otherwise, the research work in which I have placed all my hopes would suffer.[49]

The year 1955 greeted Hill with important news: the Rockefeller Foundation had decided to donate funds for the development of biophysics research at the FIA, in which physicists (Walter S. Hill), doctors (Miguel Ángel Patetta Queirolo) and biologists (Francisco Sáez) would participate. Although the funds would go to that Faculty, the agreement established with the foundation was that its administration should be in charge of a committee chaired by Hill himself. It was a donation intended for the purchase of equipment and supplies for a total of 10000 dollars.[50]

By 1957, the profile oriented towards crystallography was already a reality at the institute. During the *Fourth Rioplatense Chemical Sessions*, the institute members presented several communications. Among the names of the participants, the

[49] Hill (cited in Vaz Ferreira, 1954).

[50] Rockefeller Foundation (1955).

Norwegian Sven Furberg[51], who arrived in Montevideo as part of the UNESCO funding program in December of 1955, who during 1956 taught several courses at the university, aimed at a wide range of physicists, chemists and doctors.

In addition to this temporary incorporation that the Physics Institute managed to achieve, it is also worth noting that since 1953, when the contract with UNESCO was established, the engineer Caticha Ellis, initially trained at the Institute, was working in several European laboratories specialized in crystallography. Since that date, Ellis's bibliographic production shows a continuity of production and originality that led him to stand out in that border field between physics and chemistry, working in collaboration with the Cavendish laboratory, in Cambridge, with the University of Glasgow and with La Sorbonne. In 1957, when the term of the funding that allowed Caticha Ellis's training abroad ended, the young crystallographer wrote to his director to mention the results obtained during that time of training in Europe:

> During the tenure of the British Council scholarship in England and of UNESCO in England, Holland and France, an exploration of almost the entire field of X-ray diffraction and in particular of all the techniques that will be fundamental for the future work of the corresponding laboratory of the Physics Institute has been carried out.
>
> As a result of the work carried out, the following results were obtained:
>
> 1) - Solution of the crystal structure of benzoyl halogen peroxides. (To date the first organic peroxide whose structure has been resolved). The publication in Acta Crystallographica will be made soon.
>
> 2) - As a result of the work carried out at the Cavendish Laboratory, on techniques with Geiger Muller counters in X-ray diffraction, a work was carried out in collaboration with Dr. W. Cochran on the anomalous X-ray reflections of diamond, which may have consequences in the theory of solutions and imperfections of crystal lattices. Its title is: "The X-ray diffraction "spikes" of Diamond", and has been presented at the 4th International Congress of Crystallography to be held between July 10 and 19 in Montreal (Canada).
>
> 3) - As part of the work done in Paris, it was achieved -in collaboration with A. Rimsky- to develop a coherent theory that provides a criterion to end the process of refining an atomic structure. The theory, based on statistical considerations and on the theory of errors, finally provides a method to avoid useless calculations that -from a certain limit that we establish- lack physical meaning and greatly increase the material calculation work which is extraordinarily long in this discipline. Until now there was no quantitative criterion in this regard.
>
> (...) You are already aware of the various activities developed, which I detail more extensively in the report submitted to UNESCO, of which I will send a copy, shortly. It is clearly seen there that the program that we had proposed to develop is mostly fulfilled. However, the part related to Electron Diffraction is still pending, to which only partial attention could be paid due to lack of time or because the appropriate means were not available in the visited laboratories.
>
> As the Director perfectly knows, this technique -for its valuable applications both in pure research or applied to concrete problems of an industrial type- is very important for our laboratory, which already has the basic equipment.

[51] Sven Furberg (1920–1983) was a Norwegian chemist and crystallographer known for having predicted the helical structure of DNA, work that would later be taken up by Watson and Crick.

> Based on these considerations and the interest that Professor Wyart of the Sorbonne, Director of the Laboratorie de Mineraolgie et Cristallographie, has taken in my research, he has honored me by inviting me to stay in his laboratory until the end of the current year, and he has suggested that I request a corresponding extension of the scholarship from UNESCO. He has also kindly interceded with the authorities of UNESCO to this effect.
>
> As I consider that this extension is really in the interest of the future activities of the laboratory whose responsibility you have Trusting in the Faculty, I am writing to you to request, through your office, the necessary authorization to conduct studies on Electron Diffraction as well as to complete my previous research at the Sorbonne until the end of this year.[52]

5.5 Carlos Aragone, A Relativist in Exile[53]

The question that emerges, for the continuous and exhaustive understanding of physics at Udelar, could be posed as follows: what happened to all that training in topics related to modern theoretical physics that had been of initial interest to the Physics Institute? As we have seen, in institutional terms those old issues seemed to have been set aside in the academic trajectory of Walter S. Hill. However, it would be unfair not to note that many of the theoretical interests of physics continued to be part of Hill's concerns, even though they did not occupy his central axis of work in his last decades as a professor. As was noted, in 1948 Loedel Palumbo himself participated in academic activities at FHyC offering a course on the theory of relativity[54], a year in which Dr. Frangella, from the Institute of Radiology also offers a course on the preparation of a neutron source.[55]

As a result of the end of the Second World War and the impact of the warlike use of atomic energy, societies were involved in political discussions such as the use of atomic energy. In 1950, the newspaper El Día published several informative articles on the origins and scope of the use of atomic energy, occupying its central pages. In this context, the National Atomic Energy Commission of Uruguay was created in 1955, following in the footsteps of Argentina which had created the same national body 5 years earlier. The first Commission included: Walter S. Hill, Dr. Eduardo

[52] Hill (1957).

[53] This section marks the beginning of an ongoing investigation. The documentation on Carlos Aragone is still incipient, but we have chosen not to leave out of this story his more than significant role. Thanks to the collaboration of the Venezuelan physicists Jorge Stephany and Rita Gianvittorio we hope to coordinate actions to work on the Aragone archive at the Institute of Physics of the Simón Bolívar University in Caracas, Venezuela.

[54] The course of Loedel Palumbo included the following sections: 1. Relativity in classical mechanics. 2. The ether. 3. The Einstein's critique of space and time measurements. 4. The Lorentz transformation formulas and their elementary deduction. - Graphic representation and interpretation. - Minkowski's Universe.- 5. Relativistic kinematics. - Aberration and Doppler effect.- 6. Relativistic dynamics.- The equivalence between mass and energy and nuclear physics.- 7. Generalized theory of relativity. - The principle of equivalence.- Gravitation and the curvature of space-time.- 8. Confrontation with experience.

[55] Hill (1948).

J. Miguel, Dr. Miguel Ángel Patetta Queirolo, Dr. Vicente Elbio Sacco and Ing. Germán Villar. This commission participated in the first international meetings that were held in the following years, such as the Inter-American Symposium on the Peaceful Application of Nuclear Energy held in New York City. Hill was the main speaker on the first day of the meeting, and chaired the session on Physical Sciences and Technology, where relevant physicists such as Alvin Martin Weinberg, from the Oak Ridge National Laboratory, Walter Henry Zinn, president of the General Nuclear Engineering Corporation and who -together with Leó Szilard- discovered uranium fission by neutron bombardment; and finally Lyle Benjamin Borst, who at that time was the head of the Physics Department at New York University, but who had worked with Enrico Fermi in the past[56]. As we can see, all these names and many others that appear in the report attest to this delegation's contact with the scientific leadership in charge of the Manhattan Project.

This shows that, parallel to all the spectrography research that was being developed at the Physics Institute in Montevideo, the Physics Institute, through its director, was linked with the most relevant actors in nuclear physics at that time. However, on February 27, 1960, after 5 years of being part of the National Commission of Uruguay, Walter S. Hill submits his resignation:

> Montevideo, February 27, 1960. -
>
> Dean of the Faculty of Engineering and Surveying.
>
> Professor Engineer Enrique De Martini.
>
> Present
>
> Dear Dean: as a delegate of the Faculty of Engineering and Surveying, I have been part of the National Atomic Energy Commission since its inception in 1955. In this capacity, I have participated in most of the initiatives, projects and actions promoted by this Commission. I regret to note that the Commission's effort, developed over 5 years of work, has found little echo in official circles. In particular, our effort was directed at structuring a project for the peaceful use of nuclear energy. This project includes various aspects, such as: Law of the National Atomic Energy Commission, approval by Parliament of the Bilateral Agreement with the USA, signed in Washington on January 13, 1956, incorporation of Uruguay into the International Atomic Energy Agency based in Vienna, creation of a Nuclear Studies Center, etc. On none of these aspects has there been a favorable resolution, to date. Finally, the small subsidies from UTE and ANCAP, natural members of the National Atomic Energy Commission, have been denied for this exercise, which will make it difficult to maintain the organization of the office and the important library donated by the Atomic Energy Commission of the USA in the near future.
>
> I particularly consider that my efforts have failed despite having dedicated much time and considerable effort to them.
>
> In the future, I will not be able to dedicate the time or effort they require.
>
> For this reason, and thinking that this may favor the purposes of the Commission, I present to you my resignation from the delegation with which the Faculty honored me at the time.
>
> Without further ado, I greet you with my highest consideration.

---

[56] Hill (1957).

WALTER S. HILL.[57]

Entering the 60s, the situation of Walter S. Hill and the Institute of Physics will be altered by the changes that will mark a politically agitated decade for the country and the continent, as we will see in the next chapter. But for the purposes of training physicists, in those years, it is worth reserving a few more words for Carlos Aragone, a physicist of relativity mainly trained abroad but with significant presence in the country and the continent.

Carlos Aragone, born in 1937, joined the FIA for only 3 years in the engineering career and then continued his studies in Rome, managing to obtain his degree in mathematical sciences under the direction of mathematician Carlo Cattaneo, in 1967, with his thesis "Hamiltonian Methods in General Relativity", recognized with the *summa cum laude*. It is likely that the Italian environment to which Aragone arrived, was conducive to his interest in the famous Einsteinian theory, since Cattaneo was a disciple of Tullio Levi-Civita, a great mathematician who in the early twentieth century worked on tensor calculations, a relevant branch for the mathematical formulation of the theory of relativity. In fact, Levi-Civita maintained important correspondence with Einstein himself in the 1920s.

During his time at the FIA as a mathematics student, he passed subjects such as Introduction to General Topology (with a work called "Paracompactness") and a Functional Analysis seminar (with a monograph titled "Riesz normed spaces"), both in charge of the mathematician Juan Jorge Schaffer[58]. Aragone will complement his mathematical training working at the International Center for Theoretical Physics, in Trieste, where he takes courses on Modern Algebra, Differential Geometry, Ordinary Differential Equations, a Seminar on Partial Derivative Equations, Banach Algebra, Field Theories I, Field Theory II, Current Algebra and Computing as a Language of Physics. Between 1968 and 1970 he participates in the Latin American Physics Schools, held in Mexico and Buenos Aires.

In 1962, Walter S. Hill proposed to the FIA Council the possibility of incorporating Aragone into the teaching staff that made up the Institute of Physics. In the session where the justification for hiring Aragone, for the teaching of a short course on Relativity at Hill's proposal, Maggiolo states:

> I warmly support the proposition of Mr. Engineer Hill. I have had a conversation with the former student Mr. Aragone, and I believe that the Faculty should make an effort, in cases like this, to retain elements that have perfected themselves abroad in disciplines in which, many times, we have a shortage of teachers. It is an opportunity to form new elements, and

[57] Hill (1960).

[58] Juan Jorge Schaffer (1930–2017) was a mathematician, disciple of the first generation of mathematicians of the FIA, like Massera and Laguardia. Born in Vienna, he trained at Udelar as an engineer and then continued his tertiary education in Europe. First he obtained a doctorate in Electrical Engineering, at the Polytechnic of Zurich, and then one in Mathematics at the University of that city, in 1956. On his return to Uruguay, he taught courses at the FIA until 1968. During those years he wrote with Massera a canonical book of mathematics teaching, "Linear differential equations and function spaces" (1966). He then emigrated to the United States, where he became a professor of Mathematics at Carnegie Mellon University, until his death in 2017.

> I think that in this case, if we do not make an effort to retain Mr. Aragone, due to the very nature of the discipline, he has no other horizon than to leave the country.
>
> It occurs to me that we should offer perspectives of scientific study in the country, and a first step could be the hiring for the teaching of this short course.[59]

In 1959 and 1960, Aragone won the position of Physics professor at the State Lyceum of San Ramón (Canelones), after a competition that involved opposition of merits and five tests. After being hired by the FIA, he is assistant of Physics I (1963–1964), of General Mechanics I (1963–1966), obtains a grade one position, from 1964, for the Physics I course, and as associate professor (grade 3) for the General Mechanics I course; in 1965 he won a grade 2 assistant position at the Institute of Mathematics and Statistics; and is in charge of the Mechanics Course of the Higher Technical School of the FIA in 1966 and 1967. Finally, in 1968 and 1969, he is associate professor of General Mechanics II and Section Head at the Institute of Physics. In 1970 he is a guest professor at the Atomic Center Bariloche, by the National Atomic Energy Commission of Argentina. His last years in Montevideo, before the Coup d'Etat of 1973 that forces him to go into exile, he is professor of Theoretical Physics at the Institute of Physics, when Hill was no longer director.

His exile forces him to settle in Caracas (Venezuela), from where he will be responsible for building a regional research line on relativity and field theories. In those years he becomes an associate member of the International Center for Theoretical Physics, in Trieste (1972–1978). He is a guest professor at the Institut Henri Poincaré, of the University of Paris VI, in 1977; and accepted as Senior Research Associate, from the Physics department of Brandeis University in Massachusetts, in 1979. In 1981, he is a guest professor at the Faculty of Sciences of the University of Rosario, Argentina. In Venezuela, living in exile, he will be director of the Center for Astronomy Research in the city of Mérida, and finally director of the Institute of Physics of the Simón Bolívar University of Caracas.[60]

Carlos Aragone was a central figure in the development of the most updated orientations of relativity and field theory. He was the creator of the Latin American Regional Committee on Relativity and Gravitational Physics (COLARG), and of the Seminars on Gravitational Physics (SILARG) that began to be celebrated from their first meeting in Montevideo, in 1972.[61]. While these lines were little known in Uruguay in the 60s, they found regional receptivity from the creation of the Latin American Center for Physics (CLAF)[62]. In fact, this first SILARG, held in

[59] Faculty of Engineering (1962).

[60] Aragone to José Leite Lopes, in Lopes (1987).

[61] Gambini and Pullin (1995).

[62] The CLAF began mainly as an impulse from the Brazilian physics community, led by José Leite Lopes, with the support of UNESCO, but with relatively little participation from other Latin American physicists This situation changed over the years, and by the 70s, the community certainly had a regional and even international character. In this sense, the relationship between this Center and the International Center for Theoretical Physics, in Trieste, one of the most relevant

Montevideo, will have the presence of the new generation of physicists from the region who will develop advances in the physics of gravitational fields and relativity. In the minutes of that first meeting, the following names are found: Antonio Luciano L. Videira, Jayme Tiomno, Eliana Schnirman, Colber G. Oliveira, Ivano Damiao Soares (Brazil); Carlos Bollini, Mario Castagnino, Juan José Giambiagi, Victor M. Hamity (Argentina); Luis Alfredo Herrera, Julián Chela Flores (Venezuela); Carlos Aragone, Álvaro Restucciaa, Ricardo Siri (Uruguay), Romualdo Tabensky Rundo and Carlos Alberto López S. (Chile); and José C. Del Prado Segura (Peru). In the welcoming words written by Aragone, he acknowledges the incipient nature of this small group of Latin American researchers working on a completely new field on the continent:

> As research in Physics is reaching maturity in Latin America, its scientific community develops new structures which tend to improve the level of communication.
>
> After sponsoring first a planned activity in Solid State Physics and later in Nuclear Physics, the Latin American Center of Physics granted strong support to a meeting of the relativists (in a very broad sense) working in the area.
>
> (...) Our hope is to improve things in such a way that the next meeting could gather a greater number of relativists[63].

Aragone will be a member of the CLAF board from 1976 to 1981, years in which he will establish a friendship with the Brazilian physicist José Leite Lopes. Despite his recognized career, and having more than 75 scientific publications by the 1980s, his professional situation was very unstable. However, Aragone, who had lived a somewhat nomadic academic life, had -as a counterpoint- the possibility of understanding the reality of the various systems and scientific institutions from many parts of the world, to think about the problem of the development of physics in Latin America with the necessary critical capacity and the scientific vision of someone who anticipates future steps in his disciplinary field. Leite Lopes had sent an invitation for a Synchrotron Radiation Symposium, which had arrived late to the Astronomical Center of Mérida where Aragone was working. At that meeting, the possibility of building in Latin America a mega laboratory to attend to these lines of research in physics was discussed. Aragone responds to a letter from Leite, on January 22, 1986

> If I could have, I would have liked to hear all the points of view about how to implement a Laboratory of the magnitude that means a project of this nature, and above all the reasons to do it in this area of Physics. What I am convinced of is something trivial: our beloved governments waste so much money without thinking that implementing a "big lab" in our region, is in itself an effort with which years are gained in many directions and that the sooner they can get it working the better. If in addition the "big lab" is connected to breaking key points in the technological growth of the next 5 and 10 years, doubly better!!I have the feeling that one of the sectors that, in this order of idea, will be extremely relevant is the

partner institutions, remains open. Not only Aragone, but many actors in the Latin American physics community maintained scientific contacts with Trieste.

[63] We thank Professor Antonio Augusto Passos Videira for this documentation, belonging to the collection of his father, the Brazilian physicist Antonio Luciano L. Videira.

> one that is in an area illuminated by quantum physics and information, lasers, light technologies and condensed matter. Maybe, in our region, some of the larger countries can compete on an equal footing in this race in which although it has already lengthened, it seems to me that it still lacks more than half to finish.[64]

The correspondence between them will continue in the following years, especially on the personal topic of Aragone, who did not feel comfortable in his position at the Simón Bolívar University in Caracas and was looking for new horizons. In a new letter to Leite Lopes he accounts for this situation

> Dear Leite,
>
> I regretted that you could not be present at the inaugural session of Silarg 6. After the Silarg I had the opportunity to be in Montevideo, in the context of the recently launched PEDECIBA (Program for Development of Basic Sciences). Things are just starting to walk but very slowly. The best possible salaries in the university environment are very low compared to the cost of living ($ 700.- -) and there is no researcher career structure that complements these salaries reasonably. If they were to reach, say $ 1000.- I would take the challenge of returning to my country.
>
> Here the economic situation is a bit better. With a recent increase my salary (one of the highest in the university environment) is $770.- (14 salaries a year), but for now they yield more than the same amount in Uruguay. <u>The fundamental problem</u> is the very negative environment, even within the same university environment. (…) It is still a neutral to hostile environment. In this situation it is very difficult to develop a work in the medium and long term. Also, socially those who have come from other latitudes always feel very isolated. Caracas is a very inhumane city and neither my children nor I feel very comfortable. One lives locked up in his home. It is not an open city.
>
> At this moment I have in my charge my three younger children (Paola, 23, 4th year Industrial Engineering, Zapicán, 14, 3rd year of high school, Natalia, 12, 1st year of high school), I have been separated from their mother for 5 years, who does not live with us. In May, I turned 50.
>
> On the other hand, every time I have been in Brazil I have felt very comfortable. I notice that it is a broader and more open society and that it is very generous with those from other lands. That's why I decided to write to you to ask how you see my possibility of joining a good Brazilian scientific institution, such as CBPF or the Institute of Theoretical Physics in Sao Paulo or the University of Campinas. Even in some business of the type of CBPF researcher and also linked to CLAF partially, with some complement on its part (I have a good friendship with Giambiagi, I am also writing to him about it). At this moment my scientific interests are strings and closed superstitions and I would like to resume some work in quantum optics. I also know about the hyperinflation problems you have and maybe you are living a difficult situation. What I wanted to convey to you in this letter is fundamentally <u>what I feel</u> to hear your response. [65]

Leite Lopes tried his help in several directions, talking with the directors of the CBPF, the Institutes of Physics of Sao Paulo and Campinas, although the efforts were not fruitful. Paradoxically, as work in gravitational fields, relativity and unified theories advanced in the world, where he was clearly a reference in Latin America,

[64] Leite Lopes (1986).

[65] Leite Lopes (1987).

job opportunities were increasingly precarious or non-existent. Such was the recognition of Aragone that, on the hundredth anniversary of Einstein's birth in 1979, a celebration was held at the Center for Theoretical Studies, at the University of Miami, where an important symposium called *Orbis Scientiae* was held. It brought together the leading figures in theoretical physics from around the world to think about the possibilities and challenges of Einstein's theory of relativity and the paths to the final unification of theories. Big names like Paul Dirac, Bryce DeWitt or Abraham Pais were part of a meeting where more than a hundred relativists gathered to analyze the future of the most emblematic field of twentieth century physics. Among them, the only Latin American was Carlos Aragone [66].

5.6 Faculty of Humanities and Sciences, Hidden Tensions

The history of Physics that we are tracing forces us to go back a few decades, to the aforementioned creation of the FHyC in 1945. Making this return will allow us to show, more explicitly, a problem that appears underlying in the various areas where physics sought its development, whether at the university or national level. This problem, which is not indigenous to the Uruguayan reality and which traveled the world in various forms, was the problem of the "two cultures" [67]. With the creation of the FHyC, which sought the development of disinterested knowledge, that is, those knowledge that are valuable not in terms of their technical applicability but in their formative capacity in a humanistic sense, the emerging tension of cultures in the university context was created, now implicitly manifest.

To better understand this tension in its Uruguayan context, reading the text of Carlos Real de Azúa[68] about the Udelar, can allow us to land the problem of specialization within the context of Uruguay. In a section of the first chapter of his book *The University* (Real de Azúa 1992), which he called *Science and culture: a hostile coexistence*, the author caricatures the "two human types" that inhabit the university

[66] Perlmutter and Scott (1979).

[67] The " two cultures" was the expression that defined the emerging tension, in the societies of specialized knowledge of the twentieth century, between the literate and humanistic culture, on the one hand, and the scientific culture on the other. The idea was publicly put forward by the English physicist and novelist C.P. Snow, in 1959, in his book " The Two Cultures". But these ideas had been worked on much earlier, for example by Max Weber in his classic speech " Science as a Vocation", from 1917; or by José Ortega y Gasset, in his texts *The Rebellion of the Masses* (1929) and *Mission of the University* (1930).

[68] Carlos Real de Azúa (1916–1977) was an intellectual, professor, journalist and Uruguayan essayist. His teaching performance was in literature, as a professor at the Artigas Teachers Institute, and later as a professor of Political Science at the University. Among his most important books, we highlight his two central works of the 1960s, *The Uruguayan Patriciate* (1961) and *The Impulse and its Brake* (1964). In the case of the text we quote here, it is a posthumous edition from 1992, without specific references to the time of its writing, but that we can generally locate in his last years of life.

in this complex coexistence, which are none other than "the scientist" and "the intellectual".

> The cultural dimension at our temporal height is based on history, philosophy, religion, the arts and literature. The most rigorous and quantified sciences; mathematics, physics, chemistry, biology and all their combinations are at the other end of the spectrum.[69]

This understanding of the two models of intellectual work, recognized by Real de Azúa 30 years after the creation of the FHyC, was far from achieving a humanist integration in the formative sense of the human being, but on the contrary led to establishing tensions typical of a complete misunderstanding of what science, reaching the middle of the twentieth century, implied as a professional activity. Perhaps nothing better reflects this situation than the dispute initiated a few years after the creation of the FHyC, between the physicist Félix Cernuschi and the Hungarian philosopher Desiderio Papp.

5.7 Defending Physics with Philosophy[70]

On December 22, 1950, with the aim of organizing the courses for the following year, the Council of the FHyC decides the approval of the call for applicants to give the course "Scientific Thought", registered in the Degree in History. This course sought to influence "the "deficient scientific training" of History students[71]. The request to create such a chair was in charge of the philosopher Desiderio Papp Pollack, a professor of Hungarian origin who was in charge of the subject of Epistemology since 1947. Papp was also the author of a large number of books specializing in the History of Science, with much dissemination in the Spanish-speaking literature. Along with Papp's application for the position, the application of the physicist born in Uruguay - but trained in Argentina and Cambridge-, Félix Cernuschi, who held the chair of Astronomy of the same Faculty since 1949.

When the Council of FHyC was about to decide between both applications, on March 2, 1951, Professor Cernuschi sent a note establishing a strong denunciation against Desiderio Papp, based on two accusations: the first, of having faked his curriculum vitae when Papp held a position at the UNT, in 1946, where it documented that he had been a professor at the University of Vienna, information that was later checked and found to be false[72]. The second accusation from Cernuschi was related to Papp's academic quality:

> Mr. Papp's books are simply for dissemination and even if they had no errors, they could not be considered as sufficient merit to aspire to university teaching. The aforementioned books of Mr. Papp are full of false concepts and conceptions that clearly show that Mr. Papp does

[69] Real de Azúa (1992, 35).

[70] For a complete version of this passage, see Queijo Olano (2021).

[71] FHC Archive (1951).

[72] FHC Archive (1951c).

> not have a scientific background. He may perhaps have some literary training; but I do not like to opine on what I do not understand.[73]

This complaint causes a situation between both, and within the Council of the FHyC, which lasts for several months, even leading to criminal complaints in court. It is a tension that, as we will see, hides a much more complex background than the simple confrontation between two academics; it involves a regional tension in which the physics of the region was involved, not only in the defense of Cernuschi, but in the defense of physics and professional science against the tradition of intellectualist culture.

5.7.1 Regional Context

In June 1943, a military coup took place in Argentina that ended up overthrowing the government of Ramón Castillo. A good part of the academic and scientific community that held positions in the Universities began to publicly express their discomfort at the authoritarian advance on some study houses, such as the early intervention of the University of Litoral, in the province of Santa Fe[74]. Among these most relevant figures of Argentine academic life was Félix Cernuschi, who on November 15, 1943 was dismissed from his position in the chair of Astronomy at the UNT, being forced to flee the country due to political persecution. This university, created in 1914, had been founded following the model of the UNLP, seeking to mark an institutional identity based on science. In the framework of the scientific development of the UNT, physics played a very important role thanks to the fact that in 1925 the German physicist José Würschmidt was hired[75]. His performance in the creation of human and material resources for the practice of physics was determining, in general terms for the country, but in particular terms also for Cernuschi and Papp, since in the spaces created by him they managed to insert themselves professionally. Würschmidt was responsible for the creation, in 1931, of the Institute of Physics at the UNT, of the creation of the professorship of Physics in the province, as well as also of the creation of the Degree in Physics in 1943.[76] The Italian mathematician Alessandro Terracini and Cernuschi gave impetus to the development of the Institute of Physics in the country and the region, integrating its production to that of the scientific societies that existed at the time. One of the most significant

---

[73] FHC Archive (1951d, underlined in the original).

[74] Buchbinder (2010, 145).

[75] José Würschmidt (1886–1950) was a physicist born in Bayreuth, Germany. At the University of Erlangen he studied Physics and Mathematics, earning his doctorate in 1909. Two years later he began his academic career at that same university, first as a "privatdozent" and in 1916 as an extraordinary professor. In 1919 he was appointed professor of the "Handelshochschule", in Nuremberg, and in 1924 he took up as an extraordinary professor at the University of Cologne. He arrived in Tucumán in 1925, invited by Richard Gans.

[76] Tagashira (2012, 97–98).

milestones was the publication, in 1940, of the *Journal of Mathematics and Theoretical Physics*, which in its second issue featured an unpublished article by Albert Einstein[77]. Despite the fact that this small group of physicists managed to build the degrees in Physics and Mathematics at the UNT, some of the main academics of that young university did not manage to overcome the persecution of the de facto government. In November 1943, by order of the Executive Power, Cernuschi was dismissed from his position, initiating a political persecution against him that prohibited him from any other type of work in the Argentine universities.

On the other hand, in 1945 Desiderio Papp, after an extreme life ordeal that allows him to escape from Nazi persecution in Vienna, Spain and France, arrives in Argentina and manages to enter the UNT, after having had some experiences in other institutions in Buenos Aires. But the following year, in 1946, with the UNT intervened, he is dismissed for accrediting in his *curriculum vitae* that he had been a professor in Vienna. He is also expelled, for the same reason, from the Argentine Physics Association (AFA), of which he was an active member. This is a substantive point, because through the AFA a crusade against Papp's production, against his person, against his philosophy of the history of sciences, is launched by a community of physicists -and some philosophers- who do not hide their rejection of the way Hungarian professor does history of science, while questioning his moral integrity. And to understand even more how this rejection has its place, it is also necessary to understand that once Juan Domingo Perón assumed as president, his Minister of Health Ramón Carrillo -with whom Papp had a great friendship- decides to hire him to give a course on the History of Science at the Society of Medicine[78]. For this to happen, the issue of his dismissal had to be resolved, as a public teacher could not assume a new position in the State when he had been dismissed by that same State. And that government, we must remember, was being governed by who was a focus of harsh criticism and resistance by part of the Argentine scientific community.

Thus, returning again to 1950 and the FHyC, we can see that this dispute for the chair was really loaded with political background. The growing confrontation between Cernuschi and Papp was not a matter of the moment. Cernuschi publicly accuses Papp of being "a spy of Perón's government", placing in the middle of the discussion between both the political conflict that the Argentine community of physicists had maintained with the Argentine leader. But in addition to the political tensions, there were philosophical divisions between both.

The second accusation of Cernuschi to Papp refers to his academic quality and to what, in the eyes of the physicist, should be the necessary training to be able to assume the chair of the course "Scientific Thought". For Cernuschi, Papp's work is "simply for dissemination", and his books "could not be considered as sufficient merit to aspire to university teaching". As can be observed, Cernuschi's questioning is that of the *standard* academic, which should not only be expected for Papp, but for any academic in the area of knowledge that it is. But what is the physicist's

---

[77] Luciano (2020).

[78] FHC Archive (1951i), Kohn Loncarica (1994, 205).

criterion for deciding on the best *standard* university? For Cernuschi, the model criterion is that of his own academic trajectory and that of his colleagues.

Cernuschi trained as an engineer in Buenos Aires and, recommended by physicist Enrique Gaviola, travels to England to get a doctorate in physics. Once accepted as a doctoral student at Magdalene College in Cambridge, he begins a line of work in statistical physics, which he would have learned directly from Paul Dirac. In those years the probabilistic techniques were used to describe particle systems trying to construct the general laws of the quantum world[79]. This use of mathematics for the resolution of physics problems that links the two great phenomenal dimensions of the world that were strongly investigated since the 1930s, in the most important physics centers in Europe and United States, will be useful in his later approach to the world of astronomy. The training received by Cernuschi is that of a professional scientist, entirely organized and oriented to the resolution of problems that have to do with the understanding of the matter, its composition and behavior, but from tools and strictly defined techniques. In those years of training in England he produces a series of works that were published in the *Mathematical Proceedings of the Cambridge Philosophical Society* and that mark his initial lines of research[80]. In 1939 he publishes a paper in *Le Journal de Pysique et le Radium* on neutrons that continues this line started in England[81], and that allows him to get in contact with the chemist Moïse Haïssinsky, from the *Insitute de Radium* that had been directed by Marie Curie[82]. Back in Argentina, his production decidedly turns towards Astronomy. Thus, the study of neutrons is taken as the axis for understanding the nature of supernovas, in articles that will appear in 1939 in the journal *Physical Review*[83], at a time when Cernuschi was working with Gaviola at the National Observatory of Córdoba.

On the other hand, Desiderio Papp does not produce for the academic publishing world of journals, but does so as a writer. He does not produce articles, he produces books. He is not peer-reviewed, he is "reviewed" by readers. Even so, his books are not exempt from references to authors, concepts and terms typical of disciplinary development. His erudite knowledge about epochs, explanatory models, instruments of each period and more archaic forms of calculation on the position of the stars in the sky seems undeniable. But that does not seem to be the field in which modern theoretical physics, based on mathematics, began to think about the world of astronomy from the beginning of the twentieth century. There is between them an insurmountable gap, which is summarized in one of the interventions that Cernuschi sends to the FHyC Council on the reasons for his application for the position:

> (...) I decided to apply for the corresponding competition, not only because the subject of said course has interested me and interests me deeply (to understand scientific thought, it is

[79] Isaacs (1996).

[80] Cernuschi (1936, 1938).

[81] Cernuschi (1937).

[82] Giambiagi and Giambiagi (1996).

[83] Cernuschi (1939a, 1939b, 1939c).

essential to have worked in some of the special branches of science and especially in the most evolved ones such as the so-called exact sciences) but also so that the C.D. [Board of Directors] had the opportunity to seriously consider the problem of reviewing and investigating all the background of Mr. Desiderio Papp.[84]

Cernuschi argues that in order to speak properly about issues related to the History of Science, one must not only practice the sciences but their most advanced manifestation, the exact sciences. In other words, the history of science must be done by those who have done science. But again, this firm and structured positioning does not emanate from the debate provoked by the call; it also has a history and a context that allows us to recognize that this tension harbored two well-marked philosophical positions in the region.

In 1945, Papp publishes his work *Philosophy of Natural Laws*, a book that will provoke a tension in the regional scientific community about how to understand the History of Science. In the magazine *Science and Research* (publication of the Argentine Association for the Progress of Science) of that year appears a review by an Austrian philosopher named Hans. A. Lindemann, a pioneer in bringing the ideas of logical empiricism to Latin America. Lindemann had been invited to write an article in the first issue of *Minerva. Continental Philosophy Magazine*, a publication created by Mario Bunge, a physics student. In *Science and Research*, Lindemann reviews Papp's book where he marks his differences, in passages like the following,

> For the student and the non-specialist scientific researcher, who wants to orient himself on the problem of the book, this means rather a danger because instead of giving exact definitions, it contains many poorly defined and sometimes contradictory concepts that unsettle the unprepared reader instead of giving him clarity and precision on the majority of the philosophical and epistemological problems treated. Everything purely physical, rural, is treated with undeniable mastery and knowledge.[85]

Papp responds to this criticism, in the same *Science and Research*, in the issue corresponding to December of the same year. The basis of his response is to restore metaphysics as an element unavoidable in scientific inquiry.

> Lindemann should know that any proposition that is adopted about the relationship of the idea and the object, inevitably leads to a metaphysical thesis that is forcibly hidden in all scientific propositions. A science that eliminated all metaphysics, would renounce the reality of the objects it investigates and could not admit that the laws of nature are knowable by the human spirit.[86]

Beyond the criticism of Lindemann or the response of Papp, what interests us to highlight is that the reception of Papp's work, criticized by Cernuschi at the instance of the competition, is part of a larger cultural background, where the scientific community of the region sought to position itself against traditional academic intellectualism. For example, in a letter that the Austrian physicist Guido Beck, who worked at the National Observatory of Cordoba and who at that time was Bunge's tutor,

---

[84] FHC Archive (1951e).

[85] Lindemann (1945, 461).

[86] Papp (1945, 580).

writes to José Würschmidt, he comments on this diatribe that occurred in *Science and Research*, especially noting the questionable attitude of the author towards Lindemann's criticism

> (...) much worse is Papp's response to Lindemann's criticisms. (S. and I. December 1945). The fact that Papp is annoyed by the criticisms is humanly understandable, but a scientist must be able to abstract from that. If an author has very good factual reasons to complain about a criticism, he can complain to the responsible editorial office, can demand more factual criticisms in the future and possibly a second better criticism from another independent critic. In no case should he degrade a magazine to offer readers a dispute between an author and his critic about whether the criticism should be "good" or "bad". In addition, Mr. Papp does not try to adopt a factual position on the criticism, but simply states that the criticism "is not justified at any point". He wants to counteract the criticisms by trying to deny Mr. Lindemann's factual abilities. This is completely unacceptable and goes far beyond the scope of the problem raised. It harms the legitimate interests of the critic, the magazine, and the readers.[87]

In subsequent years, we can see that the resistance of the community scientific to the figure of Papp had not lost intensity. In 1949, in a letter addressed to Cernuschi, Lindemann maintained his appreciations about the work of the Hungarian philosopher.

> I frankly tell you that I don't have much faith in Papp's physical knowledge either, despite not being a physicist I understand something due to Schlick's classes and other physicists in Vienna. Just to not offend Rey Pastor too much, who protected Papp so much, I praised his physical knowledge. Unfortunately, it is almost always people of a lower category that Dr. Rey Pastor is protecting. I don't know why![88]

Another name comes into play here: Julio Rey Pastor. Julio Rey Pastor expressed a very different position to the one highlighted by Cernuschi and Beck, and this meant -in a way- a strong support for Papp's production. It is important to highlight the weight that Rey Pastor had in the nascent scientific community of the Río de la Plata.[89] Cernuschi himself has remembered the figure of Rey Pastor as a great mentor in his university life, the first great teacher who pushed him in scientific life[90]. But the truth is that Rey Pastor also had a career as a historian of mathematics, a facet developed from his visit to Berlin and Munich, where he came into contact with the production of Spanish mathematicians of the sixteenth century. In this sense, his sensitivity towards the History of Science may explain his understanding towards the type of production of Papp, moving away from those hard positions that characterized many of his students, such as Cernuschi himself. And a significant point of the discussion about the quality of *Philosophy of natural laws* must contain in its analysis the fact that the prologue of this work was in charge of Rey Pastor himself. In a letter that Rey Pastor writes to Guido Beck, where he mentions his "personal" problems with the AFA, he also devotes a paragraph to Papp's production,

---

[87] Videira and Puig (2020, 236).

[88] FHC Archive (1949).

[89] Ortiz (2011).

[90] Cernuschi (1994, 39).

showing the usefulness of this type of philosophy for the development of science education.

> I left aside in my previous [letter] the Papp issue; I sincerely believe that you are too rigorous with him, it is obvious that his place would not be in an Observatory or experimental chair, but it is a pedagogical Institute, of formative mission and I believe that for this purpose it is more effective than the cumbersome formulating to give the fundamental ideas about the structure of the world, and to study the history of Astronomy, interesting and dramatic. In the other institutes of secondary professorship there is an analogous chair, which seems more to judge by its performance, destined to make hate the science of the stars by force of series and trigonometric formulas.(...) I am convinced that Papp's performance will be infinitely more capable, leaving aside his enormous superiority, as culture and general formulation of all order on the average environment of Argentine university students.[91]

However, Beck's position did not seem to change. In the same letter to Jorge Wurschmidt, Beck comments on an article by Papp appeared in *Science and Research*, about "The 50th anniversary of the discovery of X-rays", dedicated to the memory of its creator Wilhelm Röntgen[92], saying that "it is not very cheerful and contains many more words than content. This is the worst South American tradition"[93]. It is important to reflect on this last expression of Beck, because it allows to introduce one of the most interesting issues for this group of physicists, such as eradicating forms of intellectual activity that they considered empty. This is the "worst tradition" that, for example, Bunge and Lindemann sought to sentence in philosophy from the few numbers appeared in *Minerva*. Bunge himself will remember in his academic memoirs the case of Papp as a case of scientific fraud, showing how that episode marked a division between the incipient community of physicists and a South American philosophy established in universities, which based their studies and courses on German-speaking authors or philosophers like Ortega y Gasset[94]. This allows us to see that the *standard* and criterion that this group of physicists sought to impose was not exclusively reduced to the sciences they practiced, but was understood as a form of intellectual life on scientistic precepts.

5.7.2 Empiricism and Rationalism, The Philosophical Dispute

The way to establish how to practice academic and scientific life, in ethical and professional terms, was what divided the waters between Cernuschi -and the physicists- on one hand, and Papp -with the support of Rey Pastor- on the other. This distance was also expressed on the philosophical ground, where both assumed positions of irreconcilable nature, which can be observed from the discussion raised

[91] Videira and Puig (2020, 250–251).

[92] Papp (1945, 491–492).

[93] Videira and Puig (2020, 236).

[94] Salmerón (1991).

about the book *Philosophy of natural laws*. The thesis of the book can be quite provocative if one is a physicist working in the first decades of the twentieth century, when so many advances were made in the description of the subatomic world especially to from the construction of experiments. For Papp, and this is established already in the first pages of the first chapter, *The law and space*,

> Science never -despite what the manuals say about the empirical bases and inductive methods of Galilean physics and therefore of current physics-, never forgot its Euclidean origins, never its spirit fundamentally betrayed its geometric ancestry. Throughout its history, it retained a nostalgia, more or less manifest, of its Hellenic and geometric lineage.[95]

Cernuschi openly expresses his rejection of this premise, in the arguments he presents to the Council of the FHyC, about the quality of Papp's production.

> (...) the fundamental thesis of his book: it revives a rationalism already overcome long ago. As is well known, for Descartes and Leibniz physics was an extension of mathematics. Descartes strove to despise experience and replace the world of physics with a world "that is the object of geometric speculation". Hegel considered that science could become a great tautology that awaited an explanatory principle from "somewhere else". We could continue mentioning a large number of philosophers who developed different aspects of rationalism. What I find novel are the contradictory and nonsensical arguments that Mr. Papp uses to "defend" an abandoned thesis.[96]

For Cernuschi "the relationships required by the knowledge of the physical world, cannot be 'deducible' from the demands of our reason". For this, he uses examples that show that the things that are, are not explained by the necessity of deductions of thought: such as that our planet has only one satellite, and not two or three; or that Newton's law of gravitation has its form by necessity of reason. Cernuschi's point is that the laws of the physical world do not show any necessity, they are synthetic a posteriori judgments whose value is "probabilistic".

> To maintain the opposite, as Mr. Papp does, it is necessary to ignore the genesis and verifiability of scientific thought, the history of science, and the elementary notions of epistemology.[97]

A second point of discussion, which derives from this great dispute about how scientific knowledge is effectively constructed, and which probably better explains the points of disagreement with Lindemann already mentioned, has to do with the philosophy of science and the understanding of science that is constructed from the precepts of logical empiricism. This point of dispute is manifested in the second chapter of the book, titled *Are natural laws, laws of nature?*. Papp intends to show, as already stated, the existence of a metaphysics that is necessary and underlies all the series of theories and laws that have been produced in the field of physics -especially- since the third decade of the twentieth century.

> Unlike philosophical thought, scientific thought admits -in its eagerness to forge an intelligible image of the phenomena that present themselves- more or less consciously three

[95] Papp (1945, 18).

[96] FHC Archive (1951h, 79).

[97] FHC Archive (1951h, 79–80).

> metaphysical hypotheses. It accepts the reality of the external world, and is convinced in advance that the incoherence in the space of phenomena, and the disorder of their succession in time, hide laws; finally, it assumes that the laws are rational, knowable by the spirit.[98]

Subsequently, Papp proposes to discuss the philosophical bases of the Vienna Circle, recognizing the authority of Ernst Mach for that current and in the attitude of physicists like Heisenberg, Born and Jordan who would carry out a "physics of the observable". The problem for Papp lies, precisely, in the reduction of what he calls the "being" to the "complex of our perceptions, the physical real to the psychic content of our consciousness". Thus, according to his view, the so-called positivism "vaporizes all the positive into an appearance without reality" and "makes of the visible and palpable nature a mere fabric of our representations". The most immediate and radical consequence that should derive from this doctrine, according to Papp, is a fall into solipsism,

> "of the 'I' forever enclosed in the narrow prison of perceptions, unable to know anything about the world around us. Such sterility would be the price paid for having eliminated from science all hypothesis, all mental construction not susceptible to observation, the fatal reward for having expelled metaphysics".[99]

Cernuschi responds to these lines saying that it is "impossible" to recompose a "more clumsy caricature of what is the "Vienna Circle"or "logical empiricism". And to discredit the positions defended by Papp he resorts to the use of a quote of authority such as Arthur Eddington, who is not a representative of the Viennese movement but who in his book *Philosophy of Physical Science* defends the empiricist modes of knowledge.

The exchange between both continues from the analysis that Cernuschi makes of each example that Papp has poured into the book, although the discussion always lies in the eternal dispute that is found between radically rationalist positions versus the radically empiricist ones. One of these controversies about how to understand the episodes of the history of science, lies in defining whether Galileo proceeded in the construction of his knowledge following deductive reasoning or did so from the induction of various scientific experiences. Again, between both there is such an absolute radicality that in certain passages of the consulted documentation, it seems unlikely to be able to establish, as surely both intended at the time, some point of definitive resolution on the opposition of readings.

The Council of FHyC of the Udelar made the decision to give the chair to Papp. The Commission that advised in this case, made up of Dean Jiménez de Aréchaga (a constitutional lawyer who, thanks to the repercussions of this case -among other problems- would resign in 1951), and the biologist Clemente Estable, did not reach an agreement. The Council had its discussion, basically centered on the first point of Cernuschi's complaint: the moral suitability of Papp, his "imposture" and the questionable relevance of this past fact in the contest that 5 years later was being carried out. Papp won, with five votes in favor and two abstentions. Cernuschi did

---

98 Papp (1945, 64).

99 Papp (1945, 64–65).

not get any votes. This could lead us to think that, in a way, this small academic battle left Papp as the winner. But the truth is that this has not been the case, if one looks from the current perspective.

The question arises, inevitably: how was it that, having won Papp the contest for the chair of Scientific Thought, his production failed to find some kind of manifest development in students or young professors who continued his ideas? It is very likely that the fact that Papp did not reside in Montevideo, and traveled every week from Buenos Aires to take care of his classes, is of great influence to hinder the possibility of thinking about some type of academic legacy in future generations. Papp worked at the FHyC until 1963, when he is succeeded by the philosopher Mario Otero, who replaces him in the chairs of Epistemology and Scientific Thought. Otero was the creator, many years later, of the Department of History and Philosophy of Science, but the philosophical profile adopted did not follow the historicist interest that was seen in the production of Papp. In his inaugural speech in the chair of Epistemology, in 1964, Otero spoke of the *Current situation and functions of the Philosophy of Science*. And there he establishes the type of program that he intends to develop:

> Having as a goal the elucidation of the telos of scientific thought is the final task that should be undertaken only in the light of particularized investigations that focus on the varied aspects of concrete scientific knowledge. The reverse approach may seem very "deep" but it inevitably leads to vagueness, to rampant speculation, when not to talk about science with a capital S. The most suitable method to avoid both this nebulous approach and an opposite pedestrian particularizing scientism is to proceed to studies of increasing generality that grasp at each stage, through a rational reconstruction, the acting and theorizing of scientists. This is precisely what the philosophy of science tries to do.[100]

Later in his talk, as if trying to resume the old discussions provoked by the book *Philosophy of Natural Laws*, almost 20 years before, Otero talks about two ways of doing philosophy of science "that do not necessarily oppose each other": one of them exemplified in the work of authors like Meyerson, Duhem or Bachelard, authors who had been a reference for Papp; the other way is the one that originates with the manifesto of the Vienna Circle. It is interesting to show that for Otero it will be this last modality that should be practiced not for its empiricist theses, nor for its rejection of metaphysics or for holding the dichotomies between analytical and synthetic judgments. What the new scientific philosophy provides, as a distinctive and determining element,

> [is the] very way of doing philosophy of science based on dialogue and criticism; it is about the collective way of doing philosophy.(…) In our view, it is precisely the use of these means of communication and philosophical dialogue, often overlooked, what constitutes the fundamental and characteristic element of the studies of philosophy of science today.[101]

The dispute over the chair of Scientific Thought, far from being a mere administrative struggle or academic egos (which it also was), revealed a whole discussion

[100] Otero (2019, 26).

[101] Otero (2019, 29).

about the consolidation of scientific authority within a university and a region that tried to initiate development in that direction. At the end of his inaugural speech, Otero reserves a space to thank those who helped him in his training that, at that moment, allowed him to assume the chair of Epistemology. Among those select names is Cernuschi. In the entire speech there is no reference to who had worked in that same chair for more than 15 years, Desiderio Papp.

The dispute based on the question of university ethics, the reading of Cernuschi's arguments and his relentless hardness of action, cannot be understood as foreign to the circle of physicists of which he was part and to the philosophy that served as support for the public defense of a professional science in a South American academic environment marked by the tradition of the "literate culture". Undoubtedly, the disciplinary construction of physics in those years, and by those first professional physicists of the Río de la Plata who had mostly had contact with elite scientific environments, imposed the criteria they understood were necessary and unyielding to institutionalize science in the universities of the continent. In a 1952 article, in the *Bulletin of the AUPCC*, Cernuschi referred to the issue of universities and the conditions for the development of scientific research in them. Among many points, Cernuschi highlights

> The spirit of scientific research is a necessary condition for the flourishing of a true social morality - since one of the indispensable pillars of this must be the defense and verification of the truth.[102]

These standards that link the professional activity of science to a moral hierarchy of society, are not a mere whimsical requirement of Cernuschi. This intellectual attitude can be seen in the writings of the region's scientists, such as Enrique Gaviola or Bernardo Houssay. This moral rule for the correct action of a university scientist served as a cohesive element among the first professional physicists in the region, aspects that are less tangible than scientific interests but seem to be of similar relevance when defining an academic group. And this ethical aspect had to do not only with scientific frauds, but with the explicit positioning against governments that - like Perón's- maintained significant differences with this science that was intended to be developed in universities.

The costly reception that logical empiricism had in the university environments of Latin America, during those years, should also be highlighted. In a way, the criticism of Papp is also the criticism of what this community of physicists understood was "the worst South American tradition" of philosophy[103]. That is why this dispute also contains more general frameworks in which to shelter. It is undeniable that the vindicative vocation of the metaphysics that Papp was doing had to be an element to attack without reservation, and so they did, both publicly and in personal correspondence, the representatives of this circle of empiricist physicists.

[102] Cernuschi (1952, 8–11).

[103] Videira and Puig (2020, 236).

5.8 Félix Cernuschi in Academia

The case of his dispute with Papp within the FHyC, allows for a better inspection of the figure of Félix Cernuschi (1907–1999), because in addition to being a physicist of international recognition, he made his scientific activity compatible with the development of a thought aimed at improving conditions for science from the 40s and until his death.

Cernuschi integrates into the Latin American scientific environment through research in cosmic rays. The interest in neutrons was part of a broader line of research on cosmic rays that had been developing in Latin America since at least 1932. As mentioned, the most prominent figure in this field in those years was the Mexican physicist Manuel Sandoval Vallarta. Vallarta studied at the Massachusetts Institute of Technology (MIT), where he got his doctorate. In 1924 he was appointed professor of physics at MIT under the direction of Vannevar Bush. After defending his thesis -on the Bohr atomic model from the point of view of general relativity and perturbation theory- he obtained a Guggenheim scholarship to study in Berlin with the elite of German physics: Einstein, Planck, von Laue, Schrödinger, among others[104]. Upon his return to the United States, Vallarta became an important figure in physics. One of the reasons for this prominence has to do with his meeting with Arthur H. Compton in Mexico, when the latter was conducting a series of experiments at different latitudes (and altitudes) of the globe, with the intention of measuring cosmic rays (and knowing the nature of these rays in their interaction with the Earth's magnetic field). From this meeting, which took place in 1932, the idea of establishing several stations in different Latin American countries to study these phenomena arose, and Vallarta played a central role in the creation of a network of physicists throughout the continent to support the initiative. Félix Cernuschi was involved in this initiative, along with Ernesto Sábato, and both had a scientific stance with Vallarta for a short period in Mexico[105]. After his expulsion from the UNT, in Argentina, Cernuschi returned to Montevideo to take up a position at the Astronomical Observatory of the Alfredo Vázquez Acevedo Institute (IAVA), a modest observatory located in one of the most important public schools in Uruguay. After the end of the Second World War and the creation of UNESCO, he was appointed scientific advisor to the organization and worked at the University of Puerto Rico. By then, he had already had the opportunity to make brief stays at Harvard and to work with the astronomical community in the U.S., although his main interest was always to return to an Argentine university. In 1949 he applied for the chair of special mathematics at the UNLP, but again his political past played a decisive role in the final rejection by the President of the UNLP, although he was recommended by the expert tribunal.

In 1947, when he had already been appointed a member of UNESCO, he drafted a report on the mission of the organization at the beginning of its international

[104] Mondragón (1999).

[105] Minor (2016).

activities in the field of science and culture. In the last part of his text, made an implicit reference to the situation of authoritarianism in some regions, of which he was very aware:

> Unfortunately, the spirit of Nazism and fascism has not yet died. It even lurks, waiting for the opportune moment for its open appearance, in some governments that were characterized by their friendship and cooperation with the Axis nations. These governments hope - and promote by all means at their disposal- a third international war. They have the morbid and criminal illusion that, after the United States of America and Russia are destroyed, they can become the rulers of the world.Peace, democracy and culture are the only trinity that can save humanity, and their worship requires the fight against militarism, poisonous nationalism, intolerance, ignorance, injustice, misery, unemployment, disease, aspirations of domination and expansion and racial prejudices. It is necessary that well-intentioned men and women produce the DDT that will eradicate the microbes responsible for these social evils. I believe it was Victor Hugo who once said, remembering what humanity owed to the Italian Renaissance, that every man should have two homelands: his own and Italy. Today, given the imperative need for peace that all peoples -have, although there are still many who have not realized this-, it is necessary that every person of good will have two homelands: their own and the rest of the world.[106]

Cernuschi finally got an appointment in Uruguay. In 1950 he applied for the chair of astronomy at the newly created FHyC[107]. By then he was already director of the Department of Physics of the Faculty of General Studies of the University of Puerto Rico. In his application letter, Cernuschi gave account of his most important academic contacts around the world, mentioning them one by one: Dr. Harlow Shapley and Dr. Bart Bock (Harvard Astronomical Observatory), Dr. Lyman Spitzer (Princeton Astronomical Observatory), Dr. Luis Erro (National Observatory of Astrophysics of Tonantzintla, Mexico), Dr. Enrique Gaviola (Puerto Rico). Enrique Gaviola (National Astronomical Observatory of Cordoba, Argentina), Dr. H. Zanstra (Professor of Astrophysics at the University of Amsterdam, Netherlands), Dr. M. Minnaerd (Utrecht Observatory, Netherlands) and Dr. Hoyle (Professor of Astronomy at Cambridge). Finally, he presented the lines of research he was developing at that time:

> I collaborate with two scientific teams conducting astrophysical research: one on cosmic rays and another on micrometeorites. In addition, I continue to investigate other astrophysical topics of my specialty, such as cosmic grains and the internal structure of stars.[108]

Upon his arrival in Uruguay, Cernuschi is interviewed by the newspaper El Bien Público, which headlines the chronicle: "Félix Cernuschi, a man of science compatriot, comes to work among us". There it is worth collecting two central aspects of his philosophical view on science. Cernuschi was a scientist who played a highly relevant role in the national public discussion, a staunch defender of science and of the philosophy of logical empiricism that he knew while he was in the United States.

---

[106] Cernuschi (1947, 14–15).

[107] FHC Archive (1950).

[108] FHC File (1950, 14–15).

> But more than the positions, to which he is always called because his name has worldwide repercussion, mark his ascending career the list of works and his participation in the most serious societies and movements scientific of America. One may disagree in some aspect of the philosophical orientation that presides over the intellectual concern of Cernuschi, but, as in his recent affiliation to the Movement for the Unity of Science, which is chaired in the USA by Prof. Phillipp Frank, it is forced to recognize the substance of the men with whom he collaborates and the importance of the intellectual work they have undertaken.[109]

It is interesting to note Cernuschi's contact with Frank, one of the genuine fathers of the new scientific philosophy that will start with the Vienna Circle[110]. This contact is personal but leaves its deep traces in some philosophical ideas that we will see in Cernuschi, and that have a clear common denominator in the scientific humanism previously conceived by Frank. Both coincide, thanks to the mentioned astronomer Harlow Shapley, director of the Harvard Astronomical Observatory. Cernuschi knows him as part of a tour and stay financed by the Guggenheim Foundation, starting in 1942.

For his part, Frank's exile in the United States had not been simple. His friend Rudolf Carnap had emigrated to the United States in 1935, taking a position at the University of Chicago. Frank and his wife Hania had decided to stay a little longer in Prague. In 1938 he was invited to give a series of lectures at various American universities, financed by the Institute of International Education, and the couple travels to that country with the plan to return. The Munich agreements, held in September of that year, which granted the incorporation of Czechoslovak areas to Germany, led the Franks to stay in the United States, and that Philipp had to find employment at some university. The first option was the University of Chicago, where Carnap and Charles Morris worked and They were leading the Movement for the Unity of Science. But the philosophy of logical empiricism, now harbored by part of American pragmatism, had strong and influential detractors. A neo-Thomist tradition, of ecclesiastical roots, played a central role in the confrontation against the new scientific philosophy in general and against the possibility of opening a job for Frank, in particular. Philosophers Robert Maynard Hutchins and Mortimer Adler, representatives of an "encyclopedic" tradition, felt that the conjunction of logical empiricism with pragmatism would result in a limited, narrow philosophy lacking human perspective [111]. Both influenced Frank not getting the position in Chicago and achieved their goal. Carnap tried to find him a position at Harvard, through physicist Percy Bridgman but, upon hearing the news, it was Haarlow Shapley who intervened to get him a position at that university's observatory. Frank, then, not only entered the said observatory, but also entered the transformation of scientific education that Harvard's president, James Bryant Conant, was starting at that time [112].

---

[109] Cernuschi (1950b, 4).

[110] Holton (2017).

[111] Reisch and Tuboly (2021).

[112] Reisch (2019).

In this intellectual environment at Harvard, Cernuschi met Frank in the most productive years of his life in North America. In a way, they both coincided not only in time and place, they both contributed (in different ways, obviously) to the construction of the Movement for the Unity of Science, and both experienced the tension of confronting an encyclopedic, almost scholastic philosophical tradition.

Since he assumed his chair of Astronomy, as the episode with Desiderio Papp makes clear, one of Cernuschi's central concerns was to dispute the cultural ground against the encyclopedic tradition. From these years on, scientific production and philosophical discussions about the sciences and their role in the country's development and teaching structures became Cernuschi's main interest. Although modest changes had been taking place in the country to secure a place for science at Udelar, Cernuschi's crusade aimed for something more. Until then, it was generally accepted that the mission of universities was to train the professional elite who could lead the country. Therefore, the most prominent professions were those oriented towards professional activities, such as law or medicine, and perhaps to a lesser extent engineering. The path to integrating a scientific trend was, therefore, difficult for the few academics who were trying to promote these studies.

These academics mainly came from mathematics, with Laguardia and Massera as references; biology, with Clemente Estable; and physics with Cernuschi and Palumbo (between Argentina and Uruguay). As we saw, Loedel Palumbo -who wrote the first scientific article on relativity in South America and studied in Germany with Planck, von Laue and Einstein- came into close contact with Reichenbach during those years in Berlin and attended the first official international meeting of scientific philosophy, held in Paris in 1935, and exchanged letters with Reichenbach in which he expressed his ideas about logic and metaphysics. In Cernuschi, his contact with Frank and Conant directly influenced his reflection on the role of the sciences for Latin America. In 1952, he wrote an article published in the Bulletin of the Uruguayan Association for the Progress of Science, titled "The University and Scientific Research" (Cernuschi 1952). He began this article by quoting the words of the President of Harvard, James B. Conant:

> The President of Harvard University, Dr. James Bryant Conant, has thus defined one of the fundamental functions of a modern university: "It is our ambition to inspire Harvard students with enthusiasm for creative intellectual work and respect for the accumulated cultural treasures of the past. In this way we perpetuate knowledge for today's posterity. Knowledge must advance and be preserved. Indeed, the only way to preserve knowledge is to advance it. If knowledge ceases to expand and develop, it loses its vitality, degrades and loses relevance to the present and the future... The pursuit of intellectual adventure must be the hallmark of every university. In the future, as in the past, our professors must be active workers, expanding the frontiers of knowledge in all directions. I hope that in our university there will never be a separation between those who teach and those who do creative work. There should be no dividing line between teaching and research. Our strength in the past depended on the spirit of research that permeated our teaching, and the members of our University have realized the importance of perpetuating the ideals of higher education and advancing knowledge in their respective fields. A university is a group of hungry people -a community of teachers and students- and therein lies the real problem for the future of any

higher culture institution. Harvard's success will depend almost entirely on our ability to attract men of the highest caliber to our student and faculty bodies.[113]

In his defense of a policy of scientific research at the university, Cernuschi focused on the idea that financial support should only be guaranteed to those professors who had shown interest in the creative work of science. In other words, Cernuschi proposed the idea of a full-time program for the development of science at Udelar. It should be noted that this idea was implemented in Uruguay 6 years later, in 1958. In the following years, Cernuschi continued to play an important role in the intellectual environment of Uruguay, establishing his position in the public discussion about science and the importance of its teaching in secondary schools and universities. In 1956, a document published by the IAVA presented Cernuschi's ideas on the teaching of sciences in secondary schools, titled "Suggestions and considerations on the teaching of sciences"[114]. That same year an essay appeared in the student magazine *University Tribune* with some "Considerations on science and scientific research in Uruguay". A small group of physicists, mathematicians, and philosophers turned to the original ideas of the Vienna Circle and the Unity of Science Movement that was projected from the United States. Cernuschi accompanied the work of Mario Bunge, the Argentine physicist and philosopher, promoter of this philosophy in Latin America, and author of the book *Causality* [115], which was the first Latin American book published by *University Harvard Press* (Bunge includes Cernuschi among the people he thanks for their collaboration in the discussion of that book).

Cernuschi directed his attack explicitly at the intellectual elite, both those who operated in the government and those who resided in the academic world. The main objective of his demand was to expose and discuss the poverty of the concept of *culture* in Uruguayan society. For him, the tradition of the humanities was responsible for neglecting the place that science should occupy in society. In 1964, Cernuschi is invited by the philosopher Mario H. Otero, who had succeeded Papp in the chair of Epistemology of the FHyC, to write in the Galileo magazine, which he created within that chair. Otero, as we saw, is also a defender of a philosophy close to logical empiricism and prepares that year a special issue dedicated to the relationship between science and humanism. There he publishes an essay, titled "Humanities and Science", which is worth recovering because it presents a diagnosis of the general consideration of humanities and sciences in Latin America:

In Latin America, the humanities are generally understood as all those studies that have nothing to do with science, mathematics, and technology. It can be argued that in this case the humanities would be reduced only to certain types of literature, the arts and other normative studies.(...) It is quite common to find scholars and artists who are completely unaware of the procedures of science and technology. I do not know any scientist or engineer

[113] Cernuschi (1952, 8–11).

[114] Cernuschi (1956).

[115] Bunge (1959).

> who does not have books of literature and art that he has read at home, and who does not have interest in theater and music. I believe that, on average, scientists and engineers are closer to modern humanism than writers and artists.[116]

This diagnosis should lead to a new idea of humanism, freed from its nineteenth-century conception and reformulated in a new holistic understanding in which science should play the central role. Cernuschi named this new intellectual project "modern humanism", which resembled some of the ideas developed at the beginning of the century by the main leaders of the Vienna Circle program, such as Philipp Frank. Cernuschi takes up this idea of rethinking the relationship between science and a new humanism, quoting a phrase from Frank himself "we must never forget that metaphysics divides people and science unites them", and adding his own vision in this new relationship

> The conflict between science and humanities is completely artificial. It is indispensable to overcome it through an integration that takes into account the multiple human values that inspire and create science and the need for many of the branches that form the so-called humanities, tend more and more to apply scientific methodology; as is actually happening with psychology, sociology, anthropology, education, philosophy, philology, history, etc. To the departments or institutes of philosophy of the universities, corresponds an extremely important work in this work of cultural integration. Therefore, it is necessary for philosophy students to continue systematic studies in the basic sciences with the purpose of being able to critically analyze the scope of scientific knowledge and to have the procedures that allow the most satisfactory and concrete understanding of human problems. Reciprocally, the science student must be duly warned that many times those scientists who despise philosophy because they only want to limit themselves to the irrefutable data of experimentation and observation, often fall, without realizing it, into outdated metaphysical dogmas. It is therefore essential that the science student understands the importance of the philosophy of science in order to logically analyze the foundations of the same.[117]

The next chapter will explore some changes in the academic and institutional environment of the university, which enlarge and expand the discussions promoted by Cernuschi since the 50s. It is mainly the project of Central Research Institutes, promoted by the rector Ing. Oscar Maggiolo, what will allow us observe the behavior of the community of physicists at the university, prior to the military intervention in the country.

Bibliography

Abarzúa Cutroni, A. 2017. Universal particles: UNESCO scientific missions in Argentina (1954–1966), *Ibero-American Journal of Science, Technology and Society -CTS*, vol. 12, n° 36, 33-60.

Archive of the Faculty of Humanities and Sciences. 1949. Letter from Lindemann to Félix Cernuschi, May 22.

Archive of the Faculty of Humanities and Sciences. 1950. CV presented by Cernuschi to the chair of Astronomy of the FHC, March 3.

[116] Cernuschi (1964, 37).

[117] Cernuschi (1964, 62–63).

Archive of the Faculty of Humanities and Sciences. 1951a. Application of D. Papp, January 10.
Archive of the Faculty of Humanities and Sciences. 1951c. Minutes of the Council of the FHC, March 2.
Archive of the Faculty of Humanities and Sciences. 1951d. Letter from Cernuschi to Dean Justino Jiménez de Aréchaga, March 2.
Archive of the Faculty of Humanities and Sciences. 1951e. Letter from Cernuschi to Dean Justino Jiménez de Aréchaga, March 7.
Archive of the Faculty of Humanities and Sciences. 1951h. Minutes of the Council of Faculty of Humanities and Sciences, May 9.
Barany, M. J. 2019. The Officer's Three Names: the formal, familiar, and bureaucratic in the transnational history of scientific fellowships. In *How Knowledge Moves: Writing the Transnational History of Science and Technology*, ed. J. Krige, 254–280. Chicago: University of Chicago Press.
Bonnecarrére, E. 1948. Suggestions for the meeting of Latin American scientists to be held this month in Montevideo. *Science and Research* 9:377–380.
Buchbinder, P. 2010. *History of Argentine Universities*. Buenos Aires: Sudamericana.
Bunge, M. 1959. *Causality: The place of the causal principle in modern science*. Cambridge, Massachusetts.
Cernuschi, F. 1936. Elementary Theory of the Critical Field of a Dielectric. In *Mathematical Proceedings of the Cambridge Philosophical Society*, vol. 32, 276–280. Cambridge: Cambridge University Press.
Cernuschi, F. 1938. Contributions to the statistical theory of adsorption: I. The elementary theory of adsorption in a first and second layer. In *Mathematical Proceedings of the Cambridge Philosophical Society*, vol. 34, 392–398. Cambridge: Cambridge University Press.
Cernuschi, F. 1939a. On the behavior of matter at extremely high temperatures and pressures. *Physical Review* 56 (5): 450–455. https://doi.org/10.1103/PhysRev.56.450.
Cernuschi, F. 1939b. A tentative theory of the origin of cosmic rays. *Physical Review* 56 (1): 120.
Cernuschi, F. 1939c. Super-Novae and the Neutron-Core Stars. *Physical Review* 56 (1): 120.
Cernuschi, F. 1947. *UNESCO. A New International Enterprise for Educational, Scientific and Cultural Cooperation*. Buenos Aires: Colegio Libre de Estudios Superiores.
Cernuschi, F. 1952. The University and Scientific Research. *Bulletin of the Uruguayan Association for the Progress of Science* 2.
Cernuschi, F. 1956. Considerations and suggestions on the teaching of science. *Annals of the "Artigas" Professors Institute* I:73–96.
Cernuschi, F. 1964. Science and humanism. *Galileo Magazine* 1.
Cernuschi, F. 1994. Félix Cernuschi researcher, teacher, scientist, humanist. In *First Edition with Corrections: Félix Cernuschi Researcher, Teacher, Scientist, Humanist*. Buenos Aires: Publication of the Secretariat of Culture and University Welfare of the Faculty of Psychology of the UBA.
Creager, A. N. H. 2014. Atomic Tracings: Radioisotopes in Biology and Medicine. In *Science and Technology in the Global Cold War*, ed. N. Oreskes and J. Krige, 31–73. Cambridge: MIT Press.
Faculty of Engineering. 1952. *Minutes of the Council*. Archive, Faculty of Engineering, University of the Republic.
Faculty of Engineering. 1956. *Resolution of the Student Assembly*. Walter Hill Collection. Archive, Faculty of Engineering, University of the Republic.
Galles, C. D., and R. Rivarola. 2012. Notes for a history of physics in Rosario. In *Physics and Argentine Physicists. Stories for the Present*, ed. D. Hurtado, 157–187. Cordoba: National University of Cordoba-Argentine Physics Association.
Gambini, R., and J. Pullin. 1995. Carlos Aragone. *Physics Today* 48 (3): 117–118.
Giambiagi, M., and M. Giambiagi. 1996. *Remembering Moïse Haïssinsky*. CBPF-CS-006/96.

Hill, W. S. 1940. Absorption of fluorescence light by the fluorescent liquid. *Anais da Academia Brasileira de Ciências* 4:261–267.
Hill, W. S. 1946. Report on the Chair of Physical-Mathematical Sciences. *Annals of the University of the Republic* 156:16.
Hill, W. S. 1947. *Correspondence Sent from Walter S. Hill to Harry Miller*. Walter Hill Collection, Archive, Faculty of Engineering, University of the Republic, Box 1.
Hill, W. S. 1948. *Walter S. Hill's Motion for His Participation in the Conference of Latin American Scientific Experts*. Walter Hill Collection. Archive, Faculty of Engineering, University of the Republic, Box 1.
Hill, W. S. 1949. *Report on the Performance of Professor Walter S. Hill*. Vol. 40, 67. Archive, Faculty of Humanities and Sciences Education, University of the Republic.
Hill, W. S. 1950. *Article Published s. d.* Vol. 40, 67. Archive, Faculty of Humanities and Sciences Education, University of the Republic.
Hill, W. S. 1951a. *Correspondence Sent from Walter S. Hill to Dean Justino Jiménez de Aréchaga*. Vol. 40, 67. Archive, Faculty of Humanities and Sciences Education, University of the Republic.
Hill, W. S. 1951b. *Correspondence Sent from Walter S. Hill to Dean Berta*. Walter Hill Collection. Archive, Faculty of Engineering, University of the Republic, Box 1.
Hill, W. S. 1952a. *Notes for the Inauguration of the X-Ray Diffraction Course*. Walter Hill Collection. Archive, Faculty of Engineering, University of the Republic.
Hill, W. S. 1952b. *Correspondence Sent from Walter S. Hill to Zoel González Ruiz*. Walter Hill Collection. Archive, Faculty of Engineering, University of the Republic.
Hill, W. S. 1953a. *Correspondence Sent from Walter S. Hill to Oscar Secco Ellauri*. Walter Hill Collection. Archive, Faculty of Engineering, University of the Republic.
Hill, W. S. 1953b. *Correspondence Sent from Walter S. Hill to Leopoldo Agorio*. Walter Hill Collection. Archive, Faculty of Engineering, University of the Republic.
Hill, W. S. 1953c. *Correspondence Sent from Walter S. Hill to Harry M. Miller*. Walter Hill Collection. Archive, Faculty of Engineering, University of the Republic.
Hill, W. S. (1953d). First tests with the "Scintiscanner" built at the Physics Institute of the Faculty of Engineering. *Bulletin of the Uruguayan Association for the Advancement of Science* 2 (4).
Hill, W. S. 1957. *Letter from Stephenson Castilla Ellis to Walter S. Hill Reporting on Scientific Results*. Walter Hill Collection. Archive, Faculty of Engineering, University of the Republic, Box 1.
Hill, W. S. 1957. *Report of the Members of the National Atomic Energy Commission of Uruguay Who Participated in the Contest*. Walter Hill Collection. Archive, Faculty of Engineering, University of the Republic, Box 1.
Hill, W. S. 1960. *Resignation Letter to the CNEA from Walter S. Hill to Dean Enrique De Martini*. Walter Hill Collection. Archive, Faculty of Engineering, University of the Republic, Box 1.
Holton, G. 2017. Philipp Frank and the Wiener Kreis: from Vienna to Exile in the USA. *Studies in East European Thought* 69:207–213.
Hormaeche, E., and W. S. Hill. 1952. Report of the delegates of the Uruguayan Association to the III Annual Meeting of the Brazilian Society. *Bulletin of the Uruguayan Association for the Progress of Science* 1:15–17.
Isaacs, A. 1996. *Oxford Dictionary of Physics*. Oxford, New York: Oxford University Press.
Jordan, P. 1953. *The Physics of the XX Century*. Mexico, DF: Fondo de Cultura Económica. [From the original German version of 1938].
Kohn Loncarica, A. G. 1994. Eloge: Desiderio Papp, 21 May 1895–31 January 1993. *Isis* 85 (4): 666–667.
Leite Lopes, J. 1986. *Letter from Carlos Aragone to José Leite Lopes. José Leite Lopes Archive*. Rio de Janeiro: FGV/CPDOC.
Leite Lopes, J. 1987. *Curriculum Vitae of Carlos Aragone. José Leite Lopes*. Rio de Janeiro: FGV/CPDOC.
Lindemann, H. A. 1945. The laws of nature. Review of Philosophy of Natural Laws by Doctor Desiderio Papp. *Science and Research* 1:10.

Luciano, E. 2020. *'In Search of a Space for Intellectual Survival': A. Terracini, Racial Laws and the Stay in Tucumán (1938–1948)*. Acc. Sc. Torino Quaderni.

Minor García, A. 2016. *Crossing Borders: Scientific Mobilizations and Inter-American Relations in the Trajectory of Manuel Sandoval Vallarta (1917–1942)*. (Doctoral Thesis). National Autonomous University of Mexico, Mexico. https://repositorio.unam.mx/contenidos/100371

Minor García, A. 2020. *Encounters: Cultural diplomacy, inter-American relations, and the mobilisation of science during the Second World War*. Crossing borders, scientific mobilisations, and inter-American relations in the career of Manuel Sandoval Vallarta, 1917–1942.

Mondragón, A. 1999. Manuel Sandoval Vallarta and physics in Mexico. *Sciences* 053.

Nascimento, M. L. F. (2015). On the "Missing Letter" to Lattes and the Nobel Prize in Physics. *Science and Society* 3 (2).

Ortiz, E. L. 2011. Julio Rey Pastor, his position in the Argentine mathematical school. *Revista de la Unión Matemática Argentina* 52:1.

Otero, M. H. 2019. *Mario H. Documents from the Otero Archive*. General Archive of the University of the Republic.

Papp, D. 1945. The 50th anniversary of the discovery of X-rays. Röntgen and his memory about the discovery of X-rays. *Science and Research* 1:11.

Perlmutter, A., and L. F. Scott, eds. 1979. *On the Path of Albert Einstein*. New York: Springer.

Queijo Olano, J. A. 2021. Cernuschi vs. Papp: The Uruguayan Dispute over the History of Science. *Transversal: International Journal for the Historiography of Science* 11.

Queijo Olano, J. A. 2023. *From the Laboratory: The Origins of Physics in Uruguay*. Montevideo: Biblioteca Plural, University Editions. (Funded by CSIC, Udelar).

Real de Azúa, C. 1992. *The University*. Montevideo: Celade.

Reisch, G. A. 2019. *The Politics of Paradigms: Thomas S. Kuhn, James B. Conant, and the Cold War "Struggle for Men's Minds"*. State University of New York Press.

Reisch, G., and A. T. Tuboly. 2021. Philipp Frank. A crusader for scientific philosophy. In *Philipp Frank. The Humanistic Background of Science*, ed. G. A. Reisch and A. T. Tuboly, 1–67. Albany: State University of New York Press.

Ribeiro de Andrade, A. M., and R. P. A. Muniz. 2006. The Quest for the Brazilian Synchrocyclotron. *Historical Studies in the Physical and Biological Sciences* 36 (2): 311–327.

Rockefeller Foundation. 1955. *Correspondence Sent from Rockefeller Foundation to the Rector of the University of the Republic*. Walter Hill Collection. Archive, Faculty of Engineering, University of the Republic.

Salmerón, F. 1991. Note on the reception of philosophical analysis in Latin America. *Isegoría* 3.

Sznajder, M. 2024. Women in Analytic philosophy: past and present, University of Warsaw, February 12–14, 2024. *KRITERION–Journal of Philosophy*.

Tagashira, R. 2012. The institutionalization and development of Physics at the National University of Tucumán until the 1980s. In *Physics and Argentine Physicists. Stories for the Present*, ed. D. Hurtado. Argentina: Editorial Universidad Nacional de Córdoba, Argentine Physics Association.

Tavares, H. D., A. Bagdonas, and A. A. Passos Videira. 2020. Transnationalism as Scientific Identity: Gleb Wataghin and Brazilian Physics, 1934–1949. *Historical Studies in the Natural Sciences* 50 (3): 248–301.

Uruguayan Society of Biology and Nuclear Medicine. 2014. *History of Nuclear Medicine in Uruguay*. Montevideo: Subimn. http://www.subimn.org.uy/society/history/.

Vaz Ferreira, C. 1954. Correspondence sent from Dean Carlos Vaz Ferreira to Rector Leopoldo Lagorio. Archive, Faculty of Humanities and Sciences of Education, University of the Republic, 40, 67.

Videira, A. A. P., and C. F. Puig. 2020. *Guido Beck: The Career of a Theoretical Physicist Through His Correspondence*. São Paulo: Livraria da Física.

Chapter 6
The Idea of Central Institutes for Scientific Research

The first ten years following the end of the Second World War marked significant economic growth for Uruguay. The economy, based on import substitution, allowed the country to continue, albeit in a much more modest tone, some of the reformist advances outlined at the beginning of the twentieth century, and during this period the industrialization process in Uruguay showed some progress. In the political context, the political division between the two historical parties was still maintained: a Colorado Party that traditionally summarized the forces of the capital; its traditional adversary being the National Party, historically representative of the forces of the countryside and the interior of the country. This dichotomy established between the two parties also characterized a duality in the country models they represented, a duality that had always found the Colorado Party victorious, but by 1958 showed a substantive shift not only in the governing forces, but in the general climate of the country. It was after the victory of a leader like Benito Nardone that the National Party managed to reach the Government, and with his arrival in power, a focus on traditionalist and neoliberal ideas was affirmed in the country that led, over the next decade, to a significant increase in state authoritarianism and violent civil response. The sixties were a decade in which Uruguay was not exempt from what was happening on the continent, where an increase, year after year, of social positions marked by radicalism manifested in constant confrontations and which would trigger the Coup d'Etat by the military forces in 1973.

Udelar was not oblivious to these transformations, changes, and climates that occurred in the country. Perhaps the most distinctive milestone was the achievement, in 1951, of an expanded form of university autonomy. Until that moment, the autonomy of Udelar was restricted in the ways of electing its authorities, and the governance of its practices and decisions based on an Organic Law. But in the middle of the twentieth century, it was sought to go even further: it included the autonomy or freedom of the chair, of ideas in the context of teaching, but, above all, budgetary autonomy. This last one, without a doubt, was the most discussed

J. A. Queijo Olano, E. Mozo Meneses, *Giants Upon Our Shoulders: A Philosophical History of Physics in Uruguay*, Studies in History and Philosophy of Science 65, https://doi.org/10.1007/978-3-031-97729-9_6

meaning of the idea of autonomy for which the fight took place in 1951 and which was finally achieved. This milestone, of utmost relevance for the institutional life of Udelar, was built from the active participation of a student body that showed characteristics typical of student youth throughout Latin America. In the fifties, Udelar had a student body not only concerned about the fate of the university, but also involved in the general problems and discussions of the country and the world, which took ideological positions on regional and global events and expected from Udelar a genuine commitment to social reality. These students found in Dr. Mario Cassinoni, a former militant of the student group of the Faculty of Medicine and a candidate for rector in the 1956 university elections, someone capable of carrying forward the project of a *new university*. What did the project of a *new university* imply? Following a Latin American sentiment, universities should be thought and oriented to respond to the social realities of the countries and the most disadvantaged populations. Put from another angle, university youth sought to leave behind the paradigm of a University forming the ruling elites.

Udelar, it was understood, still did not fulfill the imperative of putting science, technique, theory, and research at the service of society. Only a new structure, driven by a new spirit, would allow the formation of teams capable of putting technique and culture at the service of society.[1] The strengthened autonomy that the university had achieved allowed the university community itself to get involved in the changes that were beginning. The fight for the approval of the 1951 law was a stage of cohesion among university students, who saw in the won autonomy the fertile ground to structure changes. Turning the university's gaze on the country's social problems was one of those changes, but even more so was addressing them from the knowledge produced by science and technique. Therefore, with Cassinoni's rectorate, the struggle to promote science at Udelar was one of the central points of debate internally. And in this debate, the FIA was the faculty where all these problems were most vividly manifested, because there the role of the students, along with prominent professors, marked the course of change towards a new Organic Law in 1958.

6.1 No One Wants to Become a Physicist

In 1960, the Institute of Physics had 16 professors. In addition to the direction occupied by Hill, there were the teachers S. Caticha Ellis, Gilberto Baldovino, Irene W. de Ewenson, Ana M. Sanna de Álvarez, Juana Fabris de Mercatini,[2] Ruben Novaro, Martin J. Texeira, Magdalena Sánchez Varela, Julio G. Genovese, Carlos de Torres, Alcibíades Maxiera, Federico Wilson, Domingo Spadaccini, Oscar González and Pedro González. Professor Sanna de Álvarez, an industrial chemist, would enter

[1] Oddone and Paris de Oddone (2010, 186).

[2] Juana Fabris de Mercatini was Head of Laboratory in 1959, having been first hired in 1951 as a collaborator of the Institute. She earned her doctorate in Natural Sciences at the University of Florence, Italy.

the full-time regime in January of the following year. A quarter of the professors at that Institute were women, which does not seem to be a minor fact for a university traditionally made up of men.

That same year Walter S. Hill traveled, through the Rockefeller Foundation, to the United States, with the objective of meeting and working in nuclear medicine laboratories in that nation. It was a new line of work that he was developing, about which he achieved a publication that gained notable circulation.[3] Before that, in January, he sent a letter to the dean of the Faculty, engineer Enrique de Martini—who, let's remember, had been his student—to address a problem that was arising in the Faculty from the concern of a group of students who began to question the content and the general teaching of the institution, and in particular physics. As mentioned, this group of students will play a central role in this decade, associated with the revolutionary changes that characterized the political process in several Latin American countries.

> (...) I believe there is another important aspect that has not been dedicated, in practice, the effort it deserves.
>
> I am referring to the coordination between subjects that are taught in the Faculty. Not coordination on paper, but coordination in fact. It is necessary that the language of the professors and the content of their lectures reveal to the student the identity or the link that usually exists between various subjects and above all, the continuity of the teaching system. The student of an application subject, for example, should feel supported by his knowledge of previous theoretical subjects and not the other way around, verify that there is a total divorce between both, or simply, that a different language is spoken. This brings confusion and disconcertment and finally the student loses his faith in studies and in his professors with the consequent discredit of the Faculty.
>
> To avoid these inconveniences I propose the creation of small "teaching coordination meetings" ("linkage boards" of professors).[4]

It should be noted that this decade led Hill to participate more actively in the Faculty's policies, contrary to what had been all his actions in the preceding decades. One of the intentions behind this proposal to generate spaces for teaching coordination, which will later be approved as "linkage boards", was to defend the teacher's ability to organize and control his courses, and immunize from the growing student pressures. These latter were questioning the absolute independence with which the teaching body planned their activities, and the work of Walter S. Hill, and of physics training was a hot point of discussion in the Faculty Council. A year later, when this body had to deal with the request for renewal of Hill's position, Mario Wschebor,[5]

---

[3] Hill et al. (1959), Hill and Obes Polleri (1964).

[4] Hill (1960a, b).

[5] Mario Wschebor Wonsever (1939–2011) was a Uruguayan mathematician, dean of the Faculty of Sciences in 1990, when it was created after the separation of the Faculty of Humanities and Sciences. In 2012, the building of that Faculty was named after him, recognizing his contribution to the institutional construction of that Faculty. Wschebor was part of that generation of young mathematicians (like Rodrigo Arocena, Rafael Guarga, Roberto Markarián, or a little older Jorge Lewowicz) who, As students of the FIA, they drive significant changes in the Udelar that materialized in the Organic Law of 1958. It is interesting to read the book by Wschebor written in 1970,

one of the main representatives of the student order in those years—and a relevant figure in university life until his death—expressed the great disagreement that existed with the direction of the Physics Institute.

On April 14, 1961, Dean Enrique de Martini initiated the discussion about the re-election of the position of director of the Physics Institute. For his part, Wschebor, as a representative of the student order in that space, indicated that while they would approve Hill's renewal, there were some considerations they wanted to put forward and which are clearly exposed in his intervention:

> We are going to vote affirmatively for the re-election of Engineer Hill as Director of the Physics Institute.- We do so in the understanding that Engineer Hill is a talented person, that the Faculty has to count on him for its teaching and research activity; but we also do so involving a set of discrepancies with the performance of Engineer Hill in the Direction of the Physics Institute (...) Firstly, we want to point out that in the physical disciplines of the Faculty there is a lack of a policy for training teachers and attracting young staff to research and teaching tasks.- We believe that this aspect is very important; to the extent that it does not permanently train young people in teaching activity, in research activity, it will not be able to propel activity in the physical sciences in a Faculty like ours, which needs it so much. And we also understand that it is indisputable that there are a large number of Engineers with a vocation towards the physical sciences, if you want because of the nature of what Engineering is.- So we see there, in the student body of our Faculty and in the origin of what is the profession we follow, a special receptivity towards this type of disciplines, which has not been properly exploited until the present moment in the Faculty.(...)
>
> Secondly, we understand that a particular aspect of this problem of training young staff refers to the way the Institute carries out other teaching activities not framed within the study plan.- The Institute carried out in the five years we are analyzing the performance of Engineer Hill a single short course, in the year 1956.- We believe that it is essential and fundamentally in the physical sciences at this moment, that the Physics Institute is dedicated, and it has not done so, to the teaching and concern for the extension of modern physics disciplines, which at this moment are totally absent in the field of research and teaching of our Faculty and which are typical functions of the Physics Institute at the current moment.[6]

Among other criticisms presented by Wschebor in his intervention, it is mentioned that the courses taught by the Institute had no relation with the Institute itself and its activities, and neither did they do so among themselves, appearing to the students as a set of things *scattered*. Finally, the publication policy of the Institute of Physics also showed no coherent organization, there was no criterion in the purchase of foreign journals (as shown by the case of the absence in acoustics journals), nor was there coordination with the central library.

The intervention of the student Mario Wschebor, summarizes two major themes to understand the state of the development of physics: on one hand, the relevance of the contents and their proper updating; on the other, the training of young physicists in Udelar and in the country. On this matter, the traditional view that has been held about the development of Physics in Uruguay has emphasized the inability to form

Imperialism and Universities in Latin America (we thank Professor Alejandro Chmiel for this information).

[6] Faculty of Engineering (1961).

specialized human resources that would allow the long-term disciplinary course. It is worth highlighting some of the responses that were generated in that debate in the FIA Council, to the effects of showing to what extent the concern of the students was founded. Dean De Martini highlights the following, regarding the training of young physicists:

> I am forced to make some comments and some clarifications in response to what the Counselor just said.- I am going to refer first to the first point.- The Counselor [Wschebor] referred to the lack of attraction on the part of the Institute of Physics for young people who wanted to enter and work in the Institute and the reason he gave is a very considerable reason.- It is evident that due to the characteristics of our profession, more or less everyone at some point in life feels attracted to the physical sciences; so it is evident and logical reasoning.- But I must say that this should not be seen as a negative factor of the Institute, because I have had the opportunity to verify that several times the Institute has called the students of the Faculty to link them to the Institute, it has proposed jobs.- Such is the case that now there are two students from the Faculty, the bachelors Saravia and Ramos, who for a year were doing work at the Institute without any remuneration from the Faculty. I must say that these two students linked to the Institute on the occasion of a call that was made to all the students of the Faculty and where I don't know if twenty or thirty or something like that enrolled.- Moreover, I must say that these two students linked to the Institute on an occasion when we agreed with Engineer Hill to show the students the Institute before starting classes and give the students an explanation of the various tasks that were performed. From that group, which I remember I attended, which I took to the Institute when I was in the first year, showing them the various techniques that were done, some of them became interested, and from there came the bachelors Saravia and Ramos, who work.- The rest of the students from that group stopped attending the Institute.- I don't know the reasons why they stopped incurring, but I suppose that it is a logical and evident reason; the student of our Faculty does not abound in time; the student of our Faculty may have a hobby, but he cannot afford to exploit that hobby, he is extremely overloaded with the number of subjects; and if he has the need to graduate he does not take an hour of time to dedicate himself to cultivating the science that he likes.- So in that year to which I am referring, the Institute did try to have links.- It also linked professionals; there is the case for example of Engineer Azzis, who worked at the Institute of Physics honorarily; he did several jobs.- After that he was hired in Costa Rica as Director of the Institute of Physics, but he is an element that worked at the Institute of Physics of this Faculty, he took his first steps here. There were two more Engineers - and I'm talking about two or three years ago.- So I note that the doors of the Institute have been open to students and professionals; that they have tried to link them to the Institute and that the balance of all this is not a greater balance in the number of people I believe that for other circumstances and not for a neglect or for a lack of intention of Engineer Hill, of not wanting to link the staff or the position that he could have adopted to be reactive that there would be a link to the Institute.
>
> On this point I could speak at length because it is a point that I have studied in detail; not that of the Institute of Physics, but the problem that exists in the country due to the little intention, let's say, of studying the physical sciences in general.- I must say that in the Institute of Teachers unfortunately the number of aspirants to be Physics teachers, males, is one or zero per year and two or three ladies.- I'm not going to say that the courses at the Institute of Teachers are good or excellent because I'm involved, I give some of them, but I think they are not bad and that the prospects offered are good.- However, there is no student from the Faculty or from outside the Faculty who is interested in taking the Physics teaching courses.- I must say more, that in the annals of the Institute of Teachers of the year 1957 there is a publication of the one who is speaking where it dramatically raised this situation, the little concern of the our youth to dedicate themselves to physical science. So I think that the observation of the Counselor lacks absolutely weight.- I admit that it may be his, but I

> think it is not true in what pertains to an observation that can be made to a Director of Institute.- Moreover, I will say that if you want to do it now regarding this Director, at this moment, and if it has been said that in other Institutes the same thing happens, I believe that the moment, in the end, is not the most pleasant to say it when it is about his re-election.[7]

To these words of the Dean, are added those of the teacher Héctor Fernández Guido, who questions Wschebor and highlights that in the Physics Institute, during those years, young engineers were being trained, like Caticha Ellis, Sanna de Álvarez or Fabris de Mercatini. Even more interesting is to go through the text to which Dean De Martini refers, about the teaching of physics in secondary education, which reaffirms the concern presented by several physicists in the preceding decades.

In the first issue of the Annals of the Artigas Teachers Institute (IPA), from 1956, an institution responsible for the training of teachers for Secondary Education, two articles stand out that reflect the widespread concern that physics training meant for Uruguay, the one mentioned by Dean De Martini, a physics professor at that Institute, and another by Félix Cernuschi.

De Martini's text introduces some more general problems, focusing on the question of what makes a young person interested in science. From this general premise, he unravels the difficulties that modern physics, loaded with new discoveries in the last decades, presents in the face of this disinterest that is glimpsed in the Uruguayan youth. Again, the expression of the "hobby" appears, a kind of disinterested vocation, which would be the central element, for De Martini, when it comes to summoning the wills of young students.

> Faced with a science so rich in achievements, which advances at a giant pace, shouldn't we expect from our young people a marked inclination for its study? On the contrary, in our high schools, there is generally no greater interest in Physics; the same happens in university education. Other orientations are preferred. Libraries, apart from textbooks, have few works on this subject. The laboratories of our educational centers, with very few exceptions, still have the character of exhibition museums of devices. In the competitions for teaching positions, in general, the number of applicants is scarce. Also, the number of students who opt for the Physics teaching courses is small. There are not many in our environment who cultivate this science, there is no "hobby" for it. There are few conferences, courses or talks on Physics topics, and when there are, attendance is very low.
>
> What is the reason for this lack of interest of our scholars in such an attractive science? Where is that innate and disinterested desire to know, the main promoter of the development of all science?.[8]

To this series of questions, and given the discouraging diagnosis presented by the author, a series of answers are outlined that involve several dimensions: on the one hand, the obsolescence of the contents that are taught in the classrooms of the Physics teaching, totally out of step with the new challenges that the discipline presented since the beginning of the twentieth century. On the other hand, and in line with the previous, the vocation that physics could provoke is hindered by the dissection between thematic areas that are presented to the student in a totally disconnected way, but also by the levels that go from conceptual definitions to industrial

---

[7] Faculty of Engineering (1961).

[8] De Martini (1956, 117).

applications, without simultaneously showing the functioning of these relationships. De Martini addresses the problem of how physics is taught, "with chalk and blackboard", without highlighting the experimental; and in turn, when the experimental is taught none of that manages to be connected with the theories and conceptual definitions. In this bleak panorama, De Martini finds space to show how there are local examples that allow to think that the teaching of physics for young students of secondary education, is possible:

> Apart from class experiences, there is a huge field for experimentation with common use elements that are within the student's reach at home. See for example chapters VI and VII titled "Physics at home", in the excellent book "Teaching Physics" by Professor Dr. Enrique Loedel. A number of experiences are described there, which cover practically the entire course, and which are perfectly within the reach of our students of the first cycle. The outstanding feature of this type of experiences is the scarce "laboratory" material that is used, as it is reduced to some nails, threads, rubbers, squares, jars, balls, ammunition, glasses, flat glasses, mirrors, lamps, screens, wires, batteries, magnets and some other easily obtainable elements.[9]

The other article by Cernuschi (which has already been mentioned), appeared in the same issue, is titled *Considerations and suggestions for the teaching of science.* The tone of the article is very different from that of De Martini. Cernuschi is a renowned professor and researcher regionally, who has visited and worked at various universities in the world, who has held positions in international organizations for the development of science. Therefore, this text shows much more general and complex concerns, refers—as he has done on more than one occasion—to the cultural and philosophical environment of the country, and how this does not allow a fruitful development of the sciences. In other words, the challenge for Cernuschi is not so much in how and what to educate in science, but in a new humanist conception where science—and not encyclopedism—guides human life in its broadest sense.

> Science teaches us the procedures to find out how things are and the laws that govern natural processes and, consequently, to be able to use nature in part for the service of man. Science is fundamentally concerned with the how and why of phenomena and natural processes. The study and appreciation of human values and of how the growing power of science should be used to increase moral, intellectual and material progress, as well as to increase the well-being and happiness, of the greatest number of members of society, should constitute, in our opinion, the basis of modern humanism. As it is easy to see, according to the ideas exposed, modern humanism must be impregnated, in order not to degenerate into mere charlatanism, with the spirit and method of science.[10]

As previously mentioned in preceding chapters, Cernuschi develops an entire philosophical thought regarding science, in addition to being a professional practitioner. This development is inspired, mainly, by the authors of a scientific philosophy, including scientists such as Bertrand Russell, Jacob Bronowski, Philipp Frank or James Bryant Conant. Therefore, it is not difficult to recognize some epistemological principles in the pages of his text, about knowing, the relationship between disciplines and the unification of scientific knowledge:

---

[9] De Martini (1956, 128).

[10] Cernuschi (1956, 74).

> Now then, how should we continue increasing scientific knowledge when the corresponding ordinate of the knowledge accumulation curve currently is much higher than the human capacity for assimilation? Scientific knowledge is organic; that is, there are links between the different scientific chapters, which increase as scientific knowledge advances. In other words, the organization of science becomes greater with the progress of science; for example: currently, the discoveries of Physics are extremely important for the biologist, while at the dawn of Biology its practitioners could completely ignore Physics.[11]

The concrete challenges that science teaching faces, according to Cernuschi, must be able to prevent in the student any hint of mnemonic resource, which incites the student to memorize something without understanding its meaning, or in the author's words, aim at "that which remains when mere results and numerical data are forgotten". For this, he proposes a plan structured in columns where all knowledge is pedagogically structured. The column of exact sciences, would be composed of a first stage where "the fundamental and elementary knowledge of mathematics, physics, astronomy, chemistry, meteorology and geology" would be taught, considered integrally and disregarding the autonomy of the disciplines. Only from the student's development achieved over the years, is progress made towards disciplinary training. The interesting thing about this proposal, is its approach to a teaching organization not from isolated subjects, but in the style of a university institute, where the contents are conceived by a Department Section that develops all the programs. As mentioned, Cernuschi's approach accounts for a different approach, much more professional, conceiving the teacher's work in a group framework and of exchange with other teachers.

Finally, Cernuschi also takes his opportunity to point out what we have already highlighted in the previous section, the difficulty of the Uruguayan cultural environment in relation to science:

> Currently, encyclopedic education not only makes no sense, but it produces negative effects on the minds of adolescents. Indeed, by requiring them a considerable effort to memorize facts and results that they often do not understand, curiosity and the desire to learn are stifled, instead of stimulated. Therefore, any teaching that aims at encyclopedism is deforming.[12]

Returning to the FIA, and to the year in which Walter S. Hill's renewal as director of the Institute of Physics was being processed, it should be clarified that although the student demands were responded to and attacked by the teachers of the Council of that Faculty, some of the questions were addressed. On October 10, 1962, the Assembly of the Faculty's Cloister dealt with the issue of the creation of a Chair of Acoustics, an optional subject for both civil and industrial engineers, suggested by Hill himself.[13] Although the Council approved its creation subject to the budgetary possibility of the incoming year, no one doubted that the proposal was of academic relevance to the Faculty. The initiative, of a spontaneous nature, can be analyzed in

[11] Cernuschi (1956, 79).

[12] Cernuschi (1956, 80).

[13] Faculty of Engineering (1962a, b, c, d).

relation to the claims that the student order had made for the reason of Hill's renewal. That same year, as a result of a leave request from Engineer Ricardo Pérez Iribarren, director of the Institute of Electrotechnics, Hill is called to replace him for the period of months that involved such absence. He is also invited that year by the Japanese Embassy to participate in an International Symposium on Molecular Structures and Spectroscopy, to be held in that country. As the funds to participate had to be guaranteed by the Faculty, Hill suggests not investing in the ticket and allocating the same to equipment for the Institute. And in addition to that, also in 1962, the physicist Joachim Büchner is hired to join the work that was being done from the Institute of Physics with the Faculty of Medicine, in nuclear medicine.

6.2 Discussions about Physics at the University

In 1965, Hill was summoned to form a Technical Commission to analyze the relevance or not of the FIA housing in its facilities an *Inter-American Program to Improve Science Teaching*, funded by the Organization of American States (OAS), a project that had further divided the waters between students and some teachers, on the one hand, and the rest of the delegation of professors participating in university government, on the other. The first chapter of Vania Markarián's book, *University, revolution and dollars. Two studies on the Cold War cultural in Uruguay of the sixties*,[14] is a comprehensive work that allows us to account for two relevant aspects: on the one hand, and in line with what was being established, the teaching of science was a problem shared by a large part of the academic community; on the other hand, the student movement strongly questioned international (mainly American) interference in the scientific life of the University, even when they had been practiced for years.

A special national situation for science must be added, regarding a law enacted in 1961 related to the creation of the National Council for Scientific and Technical Research (CONICYT), under the Executive Power of the first National Party government in the country's history. This creation was part of a set of initiatives that UNESCO sought to promote in the countries of the region, and coincided with the emergence of national councils and commissions aimed at promoting science and technology, as happened in those years in Argentina and Brazil. The CONICYT would be led by an Honorary Board "composed of eleven members, seven appointed by the Executive Power of the government and four by the Udelar who will last four years in their functions and must be people of notable scientific or technical training". This proposal was received by the Udelar as an attack on its autonomy, which led to a series of debates about the ways and institutions with which the country should organize its scientific activity.[15] Hill, being one of the scientists appointed by

[14] Markarian (2020).

[15] Jung (2013).

the government to join the Honorary Board, probably found more hostility than he had already been receiving from the student movement, the main opponent to the creation of CONICYT.

For all these elements, the impression of the Institute of Physics of the FIA, in the turbulent decade of the sixties, seems to show its director, Walter S. Hill going through political and ideological difficulties that he had never faced as a professor. As a way to add another example, also in 1962, he raised to the Council the proposal to restructure a project to create a *Center for Research and Training in the field of nuclear energy* —which, as far as there is documentation, seems to have never been executed—. The idea was verbally supported by the counselors, although in a very vague and unspecific way.[16]

A similar example is the discussion that took place in the Assembly of the Faculty, also towards the end of 1962, about the creation of a Faculty of Mathematics. The discussion in the faculty was focused, at that time, on understanding the number of *basic* content subjects required in the training of scientists. Walter S. Hill argued strongly in favor of increasing basic subjects, referring mainly to mathematics for engineering, and reducing the number of applied subjects (according to him, following what the most advanced countries were doing). His idea was that applied training could be taken by the student once graduated, who would return to the university to continue training in specific aspects of his discipline.[17]

But the most significant proposal of the decade would be carried out in 1967, with the attempt to create Central Institutes of Research in Basic Sciences at Udelar. This impulse is carried forward when in 1958 a new Organic Law of Udelar is approved. From that moment, an active group of professors and students sought to promote a series of central modifications in the university, aimed, among other things, at ending the federative structure of Udelar, promoting scientific research by displacing professional training from centrality and promoting extension activities and activities of relationship with society. The indicated moment to promote these changes occurred with the arrival to the rectorate, in 1966, of Engineer Óscar Maggiolo. With the presidency of the Colorado Óscar Gestido, who took office in 1967, a relationship of understanding between Udelar and the Government was fostered, which promised to provide the necessary conditions for these changes (finally, Gestido died at the end of that same year of his assumption of a heart attack, and Jorge Pacheco Areco took his place). In turn, to facilitate these changes, there was the document that Maggiolo elaborated as a plan of the transformations that the university had to go through, later recognized as *Plan Maggiolo*, a restructuring plan of Udelar. In March 1967, Gestido visits Udelar, where he was received in a room full of students. The purpose of this visit was to make known the university planning for the coming years, which had to have political and budgetary support from the government to be carried out. The speech of the President of the Republic,

[16] Faculty of Engineering (1962c).

[17] Hill (1962).

Gestido, shows the harmonies that existed between these actors and that—if Gestido's death had not intervened—could have allowed a more accelerated development in the institution:

> This visit means for us, apart from the great pleasure with which we carry it out, the beginning of a dialogue that I hope can be maintained during the period in which we have been elected by the people in free elections.
>
> On our part, the dialogue will not be interrupted. As the Rector reminded us, each one with his ideas carrying out his work. It is not the differences in ideas that may interrupt the dialogue. Because it is from this encounter that solutions can emerge. In freedom for all and without detriment to anyone, neither intellectually, morally, nor spiritually.
>
> It is clear that we must express here that we disagree with some of the ideas presented, especially by the student gentlemen. I am not scared of the difference in ideas; I am scared that we may be portrayed.
>
> The abyss that is separating the so-called developed countries from the developing countries, a euphemism for underdeveloped, increases day by day. Time is against us because only a few enjoy the advances of technical and scientific changes and each advance is a greater delay that plays against us. In modern times when everything should be available to everyone, we find that what makes nations progress, general prosperity does not play for certain societies. Only culture, preparation, education and freedom can make us advance on the path of national recovery. Only in this way will we align ourselves with the formidable strength of our people, of our youth. We have an unheard-of wealth in our men, in our women, that we are wasting.
>
> I am a man grateful to my country without making a question of ideologies or political ideas. I am grateful because my country allows men of modest condition to prepare to fight in life and can occupy the highest positions. I am honored by the position I am going to hold but I am not happy. The country has big problems. We are going to face a difficult time. We can disagree on the application of methods and procedures to get the country out of this situation. What we cannot disagree on is the morality with which they are applied. What we cannot disagree on is that work is being done with devotion, without any petty interest, without any smallness, without any idea of personal or family benefit. '
>
> We must ensure, according to our ideas, the general welfare of the population and within it the welfare of the most helpless. In this sense, the University has a role to play predominant. As far as we can, we will not withhold everything it may need. And we will turn to it to require all necessary advice.
>
> What awaits us in the next five years, personally, is almost a holocaust. Like those who burn alive in defense of their ideals, we are going to burn in the next years. I only wish and only ask fortune or luck that the ashes that remain serve for something and that our Uruguay moves forward improving day by day. By the path of freedom and culture that the people can express their opinions, because as long as this persists, all disagreements and all ideologies will be fruitful.
>
> This visit that I made today for my honor, honor and satisfaction, I will always remember and I am sure that by raising the points of view above the small or large differences, we will have fulfilled for future generations the function that destiny grants us.
>
> Mr. Rector: I deeply appreciate this reception and I tell you that I hope with my best goodwill that this dialogue continues.[18]

The *Maggiolo Plan* was presented in 1968, but it had previous work that involved many discussions and meetings among university students. Thus, in relation to the

[18] University Gazette (1967, 5).

development of science, the proposal that Maggiolo intended to promote had to do with developing central science institutes (Physics, Chemistry, Mathematics, Biology, etc.), in which general training in these matters and where research activities were also developed. In the framework of the preparatory sessions, towards the end of 1966 a a meeting was called between physics professors, to which Walter S. Hill was invited. It was a meeting attended by Felix Cernuschi, Caticha Ellis, Francisco Saenz, Alfonso C. Frangella (doctor, assistant by the National Atomic Energy Commission -CNEA-), Jorge Spitalnik, of the International Atomic Energy Institute of Vienna—and former professor of the Physics Institute of the fia—and, from abroad, the professor of the Department of Physics of the University of Buenos Aires (UBA), Argentina, Juan Jose Giambiagi, and the Nobel laureate in Physics in 1959, Emilio Gino Segre. Under the title *Physics at the University*, the Rector Maggiolo cited these personalities on December 14 of that year to think about the institutional development of physics in the institution.

> The problem of basic sciences is very well known by the Uruguayans. In reality our University has not yet made a stable scheme for the training of people especially capable in any of the basic sciences. This is the case in mathematics, in physics, in biology, in social sciences, in economic sciences. Mathematicians, or physicists, or economists or sociologists, are all, I would say, a byproduct of professional careers; that is, either engineers, or industrial chemists or lawyers. And it is a concern that we have had for a long time, to the point that upon reaching the Rectorate we had thought of making an effort to correct this situation.
>
> (...) I believe that the problem we would have to discuss today is whether we want to create a scheme to launch the University into the field of teaching and physical research, as a specific objective, not as it has been until now, as part of the Faculty of Chemistry or from the School of Engineering, or from the School of Humanities and Sciences, and then, how should that plan be carried out; in what way would we prepare teachers and researchers in the theoretical field and in the experimental field; which would be the experimental branches that would be reasonable to choose in a country like Uruguay, with very large economic difficulties and even, of human material. I believe that anything we wanted to do, could not produce effective results before five years, and that everything we think has to be for a University within 10 or 15 years. That will be the goal that we should set ourselves. That is, to leave our descendants a scheme of teaching and research in the basic sciences that allows them to avoid the difficulties that all the generations that have passed through our University so far.[19]

The word was then taken by Hill as director of the Institute of Physics of the FIA. It was a very schematic and brief presentation (as had been requested by Rector Maggiolo from the participants), without dwelling on qualitative assessments or personal opinions. The portrait that Hill offers about physics in his Institute, but also from the Udelar, allows us to understand the degrees of progress that existed in this discipline by the sixties.

> (...) I believe that the problem we should discuss today is whether we want to carry out a scheme to launch the University into the field of teaching and physical research, as a specific objective, not as it has been so far, as part of the Faculty of Chemistry or from the School of Engineering, or from the School of Humanities and Sciences, and then, how

[19] Hill (1966, 1–2).

> should that plan be carried out; in what way would we prepare teachers and researchers in the theoretical field and in the experimental field; which would be the experimental branches that would be reasonable to choose in a country like Uruguay, with very large economic difficulties and even, of human material. I believe that anything we wanted to do, could not produce effective results before five years, and that everything we think has to be for a University within 10 or 15 years. That will be the goal that we should set ourselves. That is, to leave our descendants a scheme of teaching and research in the basic sciences that allows them to avoid the difficulties that all the generations that have passed through our University so far.[20]

At that meeting, Walter S. Hill's intervention was assigned to refer to "Physics in the School of Engineering". In his space, he referred to presenting a *racconto* of the history of the Institute of Physics, of the resources available at that moment, of the funding difficulties and what had been developed in the laboratory. He mentioned that the institute had responded to the three basic demands of the faculty: teaching, research and consulting, and, through the brief account of each of these functions, it is possible to reconstruct the history that has been told in this book. Hill introduces the situation of his institute showing that it has 17 members, two of them full-time, who participate in the three Physics courses for the engineering degree. The Institute's budget reached 2500 dollars a year, and it receives from revenues about 15,000 dollars. What is interesting for our purpose is what Hill indicates about research in physics at the Institute:

> As for research, I must say that it has always been linked almost always to advisory topics or collaboration with other fields of activity. Sometimes the chosen topics arose simply from the scientific curiosity that a certain problem aroused; but almost always the research in our Institute originated in the need to collaborate, imposed or forced by the environment. In a first stage of our Institute, which lasted many years, our main task was to impose physics as a basic science in the School of Engineering, and as a useful science in society. In our School there was not then a unanimous opinion as to the need to dedicate a great effort to the development of physics. It was often argued in an overly extremist way, that it was about training engineers and not physicists. Fortunately the criterion has changed; it would seem that we are now at the other extreme. No one disputes the value of physics as a basic science. I do not pretend to believe that this change is simply a consequence of our preaching; rather I think that the assimilation of new ideas universally accepted, and that in essence mean assuming that engineering is a certain form of applied physical science.
>
> But in that first stage to which I referred it was necessary to show the usefulness of physics in its collaboration with the most diverse disciplines. This led us to stimulate advice and collaborations, not only with other university institutions, but with state and private entities, such as the Ministry of Industries, Livestock and Agriculture, Banco República, Ministry of Public Health, ANCAP (National Administration of Fuels, Alcohol and Portland), UTE (National Administration of Power Plants and Electric Transmissions), etc. From these links arose some interesting research topics and also the promotion of some techniques of applied physics in other fields of utility and interest.
>
> I will cite some examples: the application of radioactive isotopes, hydrodynamics, of blood circulation, phototherapy of infant jaundice, applied spectrography, applied acoustics, radioactive minerals of the country, gamma conservation of food and perishable products, etc.

---

[20] Hill (1966, 1–2).

(...) I recognize that it may seem somewhat prosaic seen from the heights of pure science. Of course, what is exposed does not mean that we oppose the development of regular research programs. Already we have one in the Solid State Physics Department of the Institute, which has been conducting systematic research for several years in that promising field.

(...) Our purpose is to develop programs as the budget allows. But in conclusion, I want to express that even if the causes that forced us to follow this line of conduct disappear and make budgetary paths viable that would allow us to work at a higher level and on topics, if you will, more sophisticated in physics, we must not forget that we operate in a developing country and that the good implementation of science requires an economy solidly based on its own achievements.[21]

This intervention vividly summarizes what the development of physics had been up to those years: an activity completely determined by possibilities. Possibilities are what guide research. This is an element that, it seems, Hill will discover over the years. In his report on the branches of physics that the Institute has developed, nothing appears about those first intentions to venture into theoretical or nuclear physics. That would have been a desirable development, but not possible. Science cannot be something above its possibilities, possibilities that are structurally conditioned from the economic, but which are also the form of legitimation in adverse cultural environments.

In the same sense, although making harsher assessments, was Cernuschi's intervention at the table on *Physics at the University.* The professor of Astronomy from the Faculty of Humanities and Sciences valued three aspects that he considered necessary for the development of physics in the country. The first followed the line of what Hill had exposed, Uruguay had to develop a physics oriented to application, and the best laboratory to do it were the state companies of the country. This would give an outlet to the few physicists who were trained in the country, which already trained few engineers as well, compared to developed countries. A second aspect was to ensure that these few physicists could also form new vocations. Cernuschi intended for young physicists to devote themselves to secondary education, but this presented a problem. The promotion system of secondary education in Uruguay rewarded the seniority of the teacher in his position more than the levels of qualification he could achieve. Cernuschi, who knew secondary education, saw that this type of promotion was totally contrary to the requirements demanded by the scientific advancement of the modern world. Finally, his last contribution proposed the need to implement a scholarship system to train students abroad, but with caveats:

(...) (...) Finally, many scholarships must be created and foreign professors invited. In this, be very careful that the scholarship holders go to study abroad according to a well-established plan. For example, a famous colleague in the Nobel Prize, Robert Hofstaedter, who runs the largest particle accelerator in the world, was a classmate of mine; and on several occasions he has told me to send him some student, some collaborator of mine to work with him. I have told him no, never. If there is someone who can go to stay in the United States, yes, because if someone goes to work on that largest accelerator in the world, we create an enemy of our development, since back in the country -in Uruguay-, perhaps he

[21] Hill (1966, 2–3).

> gets with the generosity of Hofstaedter to publish some work with him and here he is going to say that he needs an accelerator like the one in California. And then Uruguay will have to be raffled off. But, in addition, I have seen this in some South American country, that when scholarships are given with an uncontrolled "laissez faire", well-prepared physicists are produced, one thousand percent, for export, because they have gone to study topics that at this time and in these countries they cannot then study, because such an instrument is required, that the whole country would not be enough to supply it. Then they wait for the first opportunity that is made from the United States or Europe to leave. Or if they are taken into account -Giambiagi knows some cases to which we refer, of friends that we both love very much- it turns out that immediately, a cyclotron of such a characteristic is needed, because the one that is had is not enough. And that cyclotron that needs to be bought, should it be bought because the country's development needs it? Certainly not. We must, therefore, use the experiences of all to not waste money that is needed on a large scale to use it in the best possible way.[22]

What is worth highlighting from this intervention by Cernuschi, from his vision of the development of physics in Uruguay, has to do, again, with possibilities. Latin American countries operate in the latent tension of orienting their science in relation to the main research centers of the world in a race that, due to lack of resources, is predetermined to lose, or the strategy of seeking developments in applied science, less relevant in terms of an ideal of science—of physics in this case—oriented to the understanding of natural phenomena.

Both presentations are of interest to discuss, precisely, the question about the ideals of science in countries historically lagging in their economic and social development compared to the nations of the north. In the conclusions of this work we explore some implications that this problem leaves us, to think about the idea of science from the South.

6.3 The Maggiolo Plan in Latin American Physics Networks

It is known as the *Maggiolo Plan*, the plan that the rector elaborated for the restructuring of the Udelar, of which this intention to create central research institutes is only a part. As far as the development of physics is concerned, the plan aimed to establish a Central Institute of Physics, with two Departments included: one of Astronomy and another of Meteorology. But this Plan, from which we will not delve into details now, had a much larger and more ambitious reforming purpose for the entire structure of the university. And given, precisely, its reforming character, it had repercussions outside the country's borders, not only—as we saw with the table of *Physics at the University*—by summoning several scientists from the region and other parts of the world, but because in the design itself, influences and personal and academic relationships also played their role.

The first instance of the Table, in December 1966, was followed by new stages of discussion. Rector Maggiolo wrote an invitation to the Brazilian physicist José

[22] Hill (1966, 5).

Leite Lopes, creator of the CBPF and promoter of the creation of the CLAF, so that he could participate in the preparatory stages for the creation of the Central Institute of Physics. In the official missive he sent to Leite, Maggiolo listed the pretensions he had for when he arrived in Montevideo:

> The University of the Republic is at a stage where it is strongly promoting the creation of central institutes in basic sciences and especially the Institute of Physics. Therefore, it would be a great honor and pleasure for us to have your presence among us.
>
> During your stay in Montevideo, you would develop a short course of six classes or lectures on a Physics topic of your choice.
>
> Also, due to your high scientific competence, we would be very interested in exchanging ideas with you about university organization problems and the promotion of scientific research.[23]

However, unofficially, the contacts between Maggiolo and Leite Lopes had begun much earlier through the Brazilian anthropologist Darcy Ribeiro, who was living in the country at that time and had a close relationship with the rector. In April of that year, Ribeiro wrote a letter to physicist José Leite Lopes telling him about the possible creation of a Central Institute of Physics, and, on behalf of the rector, invited him to debate for a few weeks on how to carry out the project and even offered him the direction of the institute in the stages of its implementation.

> I write to you, Leite, at the request of Maggiolo, our rector and your old admirer. He asks that you examine the possibility of coming here for a week or more to give your opinion on the best way to create a Central Institute of Physics in Uruguay. He also wishes to invite you to direct the implementation of the Institute, if this is convenient for you. In case you accept, he will send a formal invitation letter and the tickets.
>
> The position of director will be remunerated, in your case, with an exceptional salary for Uruguay of 1,000 dollars. It will probably also be possible to study travel facilities to Brazil and a permit for you to bring your car. Berta[24] is telling me that you could live here very well, with half of that salary. Indeed, life is cheap and pleasant for those who like a quiet and cordial city. After these years in Rio and Paris, this may be a pleasant prospect for you and the children. I add that the rector is excellent as a person and administrator; and he has a high regard for you and is very committed to the creation of the Institute.[25]

In this way, Leite Lopes began a double correspondence with the Udelar and its project to create a Physics Institute: on one hand, an informal one through Darcy Ribeiro (and Carlos Berta), on the other, an official one with Rector Maggiolo. Leite Lopes' informal response, sent on June 4 from Rio de Janeiro, revealed some of the difficulties the physicist was going through. In the last year, Leite Lopes had been in Paris, trying to take care of his health, but he wished to return to his country which he missed: "In Paris, despite the excellent working conditions, I felt, on one hand, that I missed, a nostalgia for the things of our Latin American world, especially, Brazilian; on the other, I feared that if I agreed to stay longer, it would become more difficult my readaptation to this harsh environment, especially, for the

[23] Leite Lopes (1967a, b, c, d).

[24] Darcy Ribeiro refers to Engineer Carlos Berta, who had been dean of the Faculty of Engineering from 1951 to 1958.

[25] Leite Lopes (1967b).

children".[26] On his return to his native country, Leite Lopes had received a similar proposal from the University of San Marcos, Peru, where he was invited to work on the installation of a scientific development program for that university. Finally, that initiative did not seem to have prospered, so Leite Lopes' uncertainties, upon receiving the invitation from Uruguay, were enormous.

> As I reflected on Rector Maggiolo's invitation, a principle of despair began to accumulate in the face of the precarious conditions of scientific work, especially in Rio de Janeiro - which were never good under any government and which now, after three years, I find have not improved.[27]

The response he formally gave to Rector Maggiolo, on June 20, showed a Leite Lopes determined to come to Uruguay to fulfill Maggiolo's invitation. However, a series of political reservations worried the Brazilian physicist. Since 1964, a coup d'état had taken place in Brazil that overthrew President João Goulart. The same generates a strongly hostile environment of persecution against left-wing political parties, which are banned and many of their members are arrested or forced into exile. Although the impact of the coup—in those years—was not of such magnitude for universities and academics (as it was in Argentina and Uruguay), it is undeniable that the political uncertainty that was experienced at that time generated fears such as those expressed by Leite Lopes (Motta 2014).

> I will be most pleased to visit your University and deliver six lectures or talks on "The symmetries of the fundamental physical laws", or better, "The symmetries of the fundamental laws of Physics".
>
> I have already made the necessary request to the university authorities asking for permission to go to Montevideo. I fear, however, that my University's decision on this matter may be delayed. Also, for health reasons, I do not wish to have possible difficulties upon my return. Under these conditions, it was suggested that an official request be made, from Ministry of Foreign Affairs to Ministry of Foreign Affairs, from Government to Government, for my trip.[28]

Although these governmental negotiations were carried out from the Rector's office, to ensure the arrival of Leite Lopes, the activity that was scheduled for August 1967 was rescheduled for a later date, in March 1968. There, the program of the activity *Seminar on an Autonomous Cultural Policy for Latin America*, which covered several discussion axes, whose inputs included a large series of documents that were sent to the participants who would attend the meeting. Leite Lopes was among them, especially for the axes titled "Demographic Policy" and "Policy of Scientific and Technological Development". This documentation was studied by Leite Lopes and represented a substantive change in his stance regarding the vision that Rector Maggiolo intended to promote. In the line of informal correspondence, on March 15, he wrote a letter to his friend Darcy Ribeiro expressing his reservations about the initiative.

---

[26] Leite Lopes (1967c).

[27] Leite Lopes (1967c).

[28] Leite Lopes (1967d).

My dear Friend,

Despite my plans to see you again, and Berta, to discuss and talk over successive nights, I am forced to inform you that I do not intend to participate in the Montevideo Meeting. It may or may not be known to you, that the idea of such a meeting and one or more corresponding documents was born in Paris, in a meeting, in meetings that took place at my house, in which participated, first of all Celso,[29] Luiz Hidelbrando[30] and some Argentine scholars, and then, they were joined by García.

The meeting having been scheduled to take place in Montevideo, by those from Montevideo -it could also have been in Chile, for example- I judged that men like Celso and Luiz Hidelbrando should be invited. I don't know if they were finally invited. Anyway, on the eve of the meeting, I don't have the list of those who were invited or the list of the works. I received one on demography, another from Maggiolo and now another one from you. I must tell you frankly that Maggiolo's work, which contains an important contribution, completely ignores the contributions of Latin Americans who for the first time had the courage to address certain issues about science and technology, including the explanation of the obstacles opposed to scientific development in the Third World as due to the alienated character of the economy and not to the ignorance of government personalities. The topic is there repeated, but everything happens as if no one had ever discussed it before, with risk, and as a priority topic. I will not fail to mention names like those of Walter Oswaldo Cruz,[31] Haity Mousstaché[32] and his colleagues from the Institute, to whom I owe so many enlightening discussions. Naturally, Maggiolo could not fail to cite Ambassador Carlos Chagas[33] and I believe he should have been invited.[34]

Leite Lopes' appreciation shows a feeling of indignation towards Maggiolo's proposal. Why? It seems that the working documents that were sent to him from Uruguay and that invite to think about the problems of an *autonomous cultural policy for Latin America*, are precisely constructed not from a Latin American

[29] Celso Monteiro Furtado (1920–2004) was a Brazilian economist, educated at the Federal University of Rio de Janeiro. From 1961 to 1964 he was Minister of Planning for the government of João Goulart. After the coup d'etat, he was forced into exile, and managed to work in several universities, one of them being the Sorbonne, when Leite Lopes was in Paris.

[30] Luiz Hildebrando Pereira da Silva (1928–2014) was a Brazilian parasitologist, with significant contributions to the study of malaria from the Oswaldo Cruz Foundation (Fiocruz). A member of the Communist Party of Brazil (PCB), he trained as a doctor at the University of São Paulo. After the 1964 coup d'etat, he was imprisoned and after being released he exiled in France, where he worked at the National Center for Scientific Research (CNRS), and at the Pasteur Institute, under the supervision of François Jacob.

[31] Walter Oswaldo Cruz (1910–1967) was also a doctor trained at the Oswaldo Cruz Foundation, which had been founded by his father. He was one of the founders of the Brazilian Society for the Advancement of Science. Like many scientists who belonged to Fiocruz, he was persecuted and his Pathology laboratory was closed, and with it all his research generated in his lifetime.

[32] Haity Mousstaché (1910–1998) was also a Brazilian doctor, and also a prominent member of Fiocruz, in the area of physiology. Also a founding member of the Brazilian Society for the Advancement of Science. Towards the end of the fifties, he participated in the design of the new University of Brasilia, a project that was led by Darcy Ribeiro. After the coup d'etat he was persecuted for his socialist affiliations and emigrated to Venezuela.

[33] Carlos Chagas Filho (1910–2000) was a Brazilian doctor responsible for the discovery of Chagas disease, a parasitic ailment occurring in the tropics. In 1966, after having participated in the main milestones of the institutional construction of science in his country, he was appointed UNESCO ambassador.

[34] Leite Lopes (1968).

originality. Leite Lopes rejects that all the experience of the construction of science in Brazil, in which he himself was a responsible actor, has not been taken for the new project in Uruguay.

6.4 The Institute of Physics in the Years before the Coup d'État

In April 1967, Walter S. Hill was invited by Dr. Roberto Bastos da Costa to the General Assembly of the Latin American Center of Physics. As already mentioned, by those years important tensions were beginning in the country that had university students—especially students—as protagonists. That precise month was developing in Punta del Este was the Summit of Heads of State and Government of the OAS, which the university strongly opposed, especially the arrival of the American president Lyndon Johnson. At a time when the university was mobilizing against North American imperialism, Walter S. Hill was invited to participate in the Inter-American Cultural Council that had been created at the meeting of the Heads of State. In a letter dated July 10, 1967, he writes to the dean of the Faculty of Engineering to request leave to participate in a meeting that this Council would hold in Washington D.C. to discuss a Program of Scientific and Technological Development.[35] It is not very strange to see that Hill's position, whether out of pragmatist motivation or simple accommodation, was contrary to that adopted by the group of students and teachers who led the reformist movement that brought Maggiolo to the Rectorate in 1966. It is also not strange, then, to understand that these would be his last years at Udelar.

By the end of 1967, the proposal to create a Central Institute of Physics at Udelar was circulating among the desks of several academics, a product of that meeting that had brought together some of the most important exponents of physics on the continent and Nobel laureate Emilio Segré.

What seemed to be a consensual agreement, a year after that scientific policy meeting at Udelar, was the need to have a single Central Institute of Physics where those areas of the university where some branch of this discipline was practiced and taught could be brought together. This document not only shows the existing developments in the Faculties of Engineering, Chemistry and Humanities and Sciences, but also includes chairs such as that of *Biophysics* in the Faculty of Medicine, the General Mechanics departments, the Electronics and Fluid Mechanics departments (FIA), the Physics chair of the Artigas Teachers Institute, the Institute of Mathematics and Statistics (FIA), the chairs of Physical Chemistry (FIA) and Chemistry (FHyC). There was a nuclear reactor, belonging to the National Atomic Energy Commission, but it had never been started due to lack of resources and not having a suitable place for its installation. But the interesting thing about the summary document, which was intended to be established but never achieved its realization, proposed some actions:

[35] Hill (1967a, b, c).

3.1. The first measure to be taken, which unequivocally arises from the round table "Physics in Uruguay", is the merger of the three university centers into one, in order to coordinate efforts, promote greater contact between the different specialists and avoid useless duplications.

3.2. The second measure is to program a concrete plan, within a period of five years, for the training of theoretical and experimental physicists in the country, at the bachelor's level, and abroad at the Ph.D. level, through scholarships selected by topics and taking into account very especially the scientific level of the beneficiary (graduates, engineers, chemists, graduates of the Artigas Institute), as well as their vocation and predisposition to dedicate themselves to scientific research and higher education.

3.3. Link the training of physicists at the bachelor's level with that of high school teachers, passing the Institutes of this kind that exist in the country, to the University. This requires that the Primary and Secondary Education Entities recognize the studies that are carried out at the University, together with those of the Institute of Educational Sciences, which would be created simultaneously (teaching and research in pedagogical disciplines) for the exercise of teaching in Secondary Education, University of Labor, Magisterial Institute of Higher Studies.

This solution will allow physics teachers (as well as those of all other scientific and humanistic disciplines) to be trained in direct contact with higher education teachers and researchers of the basic disciplines, in the same way as will happen in the future with engineers, doctors, chemists, agronomists and other liberal professionals. It must be emphasized, because this is sometimes forgotten in our country, that higher education, whatever the discipline, basic or applied, that is considered, can only be carried out at a good level if the student is in direct contact with the researchers of the same discipline.[36]

At the end of 1967, Walter S. Hill sent his collaborators Caticha Ellis, Luis Saravia, Néstor Azziz and José Salzberg a form so that they could join the International Centre for Theoretical Physics, created in 1964 by physicist Abdus Salam, in Trieste, Italy.[37] By 1969, the scientific production of the Physics Institute showed its relative stagnation both in terms of its scope and the topics of development. Even so, the physicists of those years dedicated part of their hours to conducting laboratory experiments that they later published.

"Study of a yellow mineral existing in the radioactive rocks of Sauce de Pan de Azúcar using X-ray diffraction and Spectrography" by Juana F. de Mercatini and Irene W. de Ewenson.

"Anomalous X-ray diffraction" by S. Caticha Ellis.

"Composition of renal and biliary calculi by X-ray diffraction" by A.M. Sanna de Álvarez.

"Composition of Portland cements by X-ray diffraction" by A.M. Sanna de Álvarez and Juana F. de Mercatini

"Influence of the mineralogical composition of Portland cements on the physico-mechanical properties" by G. Schaich and A.M. Sanna de Álvarez

"Elementary hydrodynamic basis of an analog of the global blood circulation: congestive pulmonary failure" W.S. Hill and J.O. Polleri

"Hydrodynamic analysis test of blood circulation" W.S. Hill, J.O. Polleri and A.L. Matteo.

"On the syndrome of respiratory distress in the newborn. Congestive pulmonary failure." W.S. Hill and J.O. Polleri.

"Automatic registration of the location of a radioisotope" W.S. Hill

[36] Hill (1967b).

[37] Hill (1967c).

"Circulatory adaptation at birth" W.S. Hill, J.O. Polleri and A.L. Matteo.
"On the theoretical basis of electromagnetic methods to measure rates of flow" W.S. Hill
"Cross Scanning" W.S. Hill and J. Traibel
"Dosimetry problems in Cycloterapy" Dr. A. Frángela and W.S. Hill
"General theory of physical magnitudes" W.S. Hill
"Design and execution of a reverberation time meter" J.L. Zuccoki.
"X-ray diffraction study of artery calcifications" A.M. Sanna de Álvarez and Juana F. de Mercatini
"Manganese minerals from the Guaycurú area (San José Department) A.M. Sanna de Álvarez and Juana F. de Mercatini
"Programs for crystallographic calculations, made for the I.B.M. 1401 computer" L. Saravia, J. Salzberg and J.L. Duomarco
"Elementary hydrodynamics of congestive pulmonary failure" W.S. Hill and J.O. Polleri.
"Cross Scanning: simple way to improve resolution" W.S. Hill and J. Traibel
"Radioactive elements in tertiary and quaternary fossil bones" W.S. Hill
"R search in prematurity" W.S. Hill and J.O. Polleri
"Identification of some crystalline substances from the Ilmenite-Monazite Sands of the Uruguayan Coast by X-ray Diffraction" S. Caticha Ellis. (Hill 1967c).

Walter S. Hill wrote to Dean Julio Ricaldoni on October 28, 1969, to request a leave of absence after having started his retirement procedures. That same year Carlos Aragone taught the first course at the FIA on Field Theory.

In 1970 Walter S. Hill left his position, after 39 years of work at the Institute of Physics. However, this would not be his last participation in the institution, since, once the university was intervened by the de facto government, in 1976, he was summoned to join an advisory committee of the intervening dean Alfredo de Santiago. On that occasion, Hill was recognized as an honorary professor. Three years later, in 1979, the Faculty of Engineering still published the text "Physics Course III," authored by him, a work for students elaborated on the introduction to the physics of electromagnetism.

Some of the physicists who were part of the Institute or Udelar had to emigrate. Aragone settled in Venezuela and formed the Institute of Physics at Simón Bolívar University. Luis Saravia completed his doctorate at Northwestern University in the United States, researching semiconductor physics, and in 1974 returned to Latin America to settle in Argentina, where he began working on solar energy. Néstor Azziz earned his doctorate at Pennsylvania State University, working on theoretical physics of the nuclear field.

Bibliography

Cernuschi, F. 1956. Considerations and suggestions on the teaching of science. *Annals of the "Artigas" Professors Institute* I:73–96.

De Martini, E. 1956. Physics in the First Cycle of secondary education. *Annals of the "Artigas" Professors Institute* 1:115–131.

Faculty of Engineering. 1961. *Council Minutes*. Archive, Faculty of Engineering, University of the Republic.

Faculty of Engineering. 1962a. *Minutes of the Faculty Council*. Archive, Faculty of Engineering, University of the Republic.

Faculty of Engineering. 1962b. *Council Minutes*. Archive, Faculty of Engineering, University of the Republic.

Faculty of Engineering. 1962c. *Council Minutes*. Archive, Faculty of Engineering, University of the Republic.

Faculty of Engineering. 1962d. *Council Minutes*. Archive, Faculty of Engineering, University of the Republic.

Hill, W. S. 1960a. *Resignation Letter to the CNEA from Walter S. Hill to Dean Enrique De Martini*. Walter Hill Collection. Archive, Faculty of Engineering, University of the Republic, Box 1.

Hill, W. S. 1960b. *Correspondence from Walter S. Hill to Dean Enrique de Martini*. Walter Hill Collection. Archive, Faculty of Engineering, University of the Republic.

Hill, W. S. 1962. *Intervention of Walter S. Hill in Assembly of the Faculty of Engineering and Surveying*. Walter Hill Collection. Archive, Faculty of Engineering, University of the Republic.

Hill, W. S. 1966. *Physics at the University*. Walter Hill Collection. Archive, Faculty of Engineering, University of the Republic.

Hill, W. S. 1967a. *Correspondence from Walter S. Hill to Acting Dean Arturo Carbonell*. Walter Hill Collection. Archive, Faculty of Engineering, University of the Republic.

Hill, W. S. 1967b. *Project to Create a Central Physics Institute at the University of the Republic*. Walter Hill Collection. Archive, Faculty of Engineering, University of the Republic.

Hill, W. S. 1967c. *Letter from Hill to Ellis, Saravia, Azziz and Salzberg to Send Form to the ICTP*. Walter Hill Collection. Archive, Faculty of Engineering, University of the Republic.

Hill, W. S., and J. Obes Polleri. 1964. Elementary hydrodynamic basis of an analog of the global blood circulation: congestive pulmonary failure. In *Pulsatile Blood Flow*, ed. Ernst O. Attinger, 407–412. New York: McGraw Hill.

Hill, W. S., J. Obes Polleri, et al. 1959. *Essay on Hydrodynamic Analysis of Blood Circulation, Engineering Journal*. Montevideo.

Jung, M. E. 2013. The UDELAR and the creation of CONICYT: debates and conflicts (1961–1964). *Encuentros Uruguayos Journal 6* (1): 32–49.

Leite Lopes, J. 1967a. *Letter from Rector Oscar Maggiolo. José Leite Lopes Archive*. Rio de Janeiro: Getulio Vargas Foundation.

Leite Lopes, J. 1967b. *Letter from Darcy Ribeiro. José Leite Lopes Archive*. Rio de Janeiro: Getulio Vargas Foundation.

Leite Lopes, J. 1967c. *Letter from Leite Lopes to Darcy Ribeiro and Carlos Berta. José Leite Lopes Archive*. Rio de Janeiro: Getulio Vargas Foundation.

Leite Lopes, J. 1967d. *Letter from Leite Lopes to rector Maggiolo. José Leite Lopes Archive*. Rio de Janeiro: Getulio Vargas Foundation.

Leite Lopes, J. 1968. *Letter from Leite Lopes to Darcy Ribeiro. José Leite Lopes Archive*. Rio de Janeiro: Getulio Vargas Foundation.

Markarian, V. 2020. *University, Revolution and Dollars. Two Studies on the Cultural Cold War in Uruguay of the Sixties*. Montevideo: Penguin Random House Grupo Editorial.

Motta, R. P. S. 2014. *Universities and the Military Regime: Brazilian Political Culture and Authoritarian Modernization*. Editora Schwarcz-Companhia das Letras.

Oddone, J., and B. Paris de Oddone. 2010. *History of the University of the Republic*. In *The University from Militarism to the Crisis. 1885–1958*, vol. 2. Montevideo: Editions of the University of the Republic, University Classics Collection.

University Gazette. 1967. Information bulletin. *University Gazette* 3:14.

Chapter 7
Conclusions

7.1 Philosophical Conclusions

This history of physics from Uruguay encourages two types of reflections, connected but not necessarily the same. The first one allows us to think about what we can and should understand under the label of *Physics*, within the set of knowledge, practices and possibilities that were established in this country and the region. The second reflection has to do with that more properly philosophical question, the one that underlies every deep analysis that is provoked from the philosophy of science: what is or should we understand by science? If we argue that both dimensions are intimately connected it is because we are guided by the thought that was reflected in the introduction, reduced in Goethe's phrase: "the history of science is science itself".

Reversing these two planes of reflection, we could ask ourselves: can we give a definition of science that fits the context and possibilities of Latin America without properly knowing how science has been done in each country of our region? To think about the science cultivated in countries like Uruguay, particularly physics, it was imperative for us to fulfill the task of constructing a history that had not been carried out, at least not in a totality that would allow us to have a great image over time.

7.2 Reflections on Physics in Uruguay and Latin America

The history of physics in Uruguay, from its first steps at the end of the nineteenth century to the end of the 1960s, is a story of constant adaptation to challenges, both intellectual and scientific. The development of this discipline shows how it went

J. A. Queijo Olano, E. Mozo Meneses, *Giants Upon Our Shoulders: A Philosophical History of Physics in Uruguay*, Studies in History and Philosophy of Science 65, https://doi.org/10.1007/978-3-031-97729-9_7

from being shaped by local needs for its useful application, to becoming an institutionalized discipline, but always subject to the ups and downs of a complex context, with individual efforts and sometimes isolated for its development.

The physics of the nineteenth century, seen from the Rioplatense point of view, we understand it as a malleable concept because unlike Europe -which always officiated as a mirror of utopias- where physics was consolidated from a theoretical and experimental approach, in Uruguay the discipline emerged adapted to the philosophical, social and productive demands of a country in formation. Its introduction to the academic field was mostly through practical applications, far from a purely theoretical approach. In this way, it is not possible to study the genesis of Rioplatense physics from the classic narratives that the history of ideas has offered, regarding the period, but through what the available records show us. These records show that it is not the precepts of the mathematization of movement, the geometrization of space, or the model of experimental science that introduce the disciplinary autonomy of physics, but that it attended to various very particular contextual conditions: a very present philosophical heritage still contaminated by the scholasticism, which sought to agree with the scientific knowledge that arrived from Europe, which sought to adapt without achieving its full understanding. This process of autonomy is accompanied by the institutional transformation that supposed the creation of the Universidad Mayor de Montevideo, in 1849; and a growing need, throughout the nineteenth century, for industrial development from the applications that the new knowledge seemed to provide.

At least these aspects must be taken into account to understand this idea of a malleable physics. First, the movement of the *physicists* who sought, above all, to combat the dominant ecclesiastical formation. A stage where physics, despite being looking for new directions, is still impregnated with the approaches, terminologies and tools of philosophy. An environment where, although, they were times where new figures and discoveries like those of Newton or Faraday allow questioning the dominant philosophical formation, the general Physics programs, were closely related to the philosophical approaches and showed concern for the qualities of the phenomena and their essential meaning, which was necessary to define to acquire better explanations about Nature. Likewise, we see also that there is room in these programs for a particular Physics much more oriented to attend solutions of experiments that can provide immediate applications.

The form of scientific practice that characterized the first chairs, such as the Physico-Mathematical one created in 1849 guided by personalities like Villegas, Jacques, Du Pasquier, who present different formations and therefore imprint different conceptions of physics, sow a feeling of disconcert in relation to the specific contents for such a subject. Also, as it was built as an isolated chair, not conceived as part of a specific formation, it becomes a space lacking orientation. This last point is accentuated with the constant financing problems that occurred to be able to hire teachers, and in the concern that it was necessary to acquire specialized bibliography that would serve as a guiding model.

This malleability of the idea of physics is altered with the creation of the faculties of Medicine and Mathematics, which provided a place for physics (and especially certain areas of it, such as electricity) that was significant to give it an orienting sense that it had not had until then. The diagnostic applications of physics in the medical area promoted by Jacinto de León, or the emphasis on the electrical applications in productive machinery highlighted by Williman, are nothing but the transformation of the same idea of physics into a set of knowledge of direct and concrete application and although it began to have that halo of legitimacy as a discipline, clearly they did not have a preponderant place in the formative structure neither of doctors, nor of surveyors, nor of architects.

When can we recognize an emerging path towards the disciplinary autonomy of physics in Uruguay, which had always been oriented to technical and professional applications in the nineteenth century? Our hypothesis is that it was from certain changes in mathematical training in the engineering career, at the beginning of the twentieth century, what provides conceptual and theoretical tools to certain actors, to address the mathematized requirements of modern physics. The discussions in the Faculty of Engineering about the different geometries and algebra, promoted by Eduardo García de Zúñiga, show that from this new training a group of engineers managed to approach in a more technical way the complex physics that was being developed at that time in Europe.

In 1916, in the Faculty of Engineering, the creation of a Higher Institute of Physics was proposed, driven by figures such as Eduardo García de Zúñiga and Bernardo Kayel. Although this initiative did not materialize at the time, it marked an important step towards the institutionalization of physics in the country. At the same time, in 1913, the Institute of Radiology was founded in the Faculty of Medicine, which represented a significant advance for experimental physics in Uruguay, showing the potential of physics in practical applications and its contribution to medical development in the country. At this point, we see how physics manages to integrate into the academic discussion, and showing the two spaces that have historically been involved in the development of the discipline: engineering and medicine.

From the twenties of the twentieth century, we see a series of engineers who decidedly turn to the attention of the pure problems of physics: Amadeo Geille Castro, Enrique Loedel Palumbo and Walter S. Hill. With diverse trajectories, the three are part of a generation that changes the work on physics and encourages its autonomy. The academic and national environment opposed to the development of basic sciences, managed to become more flexible with the arrival of Albert Einstein to the country, in 1925. Despite not understanding the scientific implications of the new general theory of relativity, the cultural environment recognized the new intellectual that the world openly recognized.

Physics acquires a relative disciplinary autonomy with the creation of the Institute of Physics in the Faculty of Engineering, at the end of 1941. Without a doubt, this is a significant change, a disciplinary process of autonomization, which impacts the opening of possibilities for development in consolidated fields of physics. For example, from the creation of the Institute, physics in Uruguay begins to be conceived as a discipline integrated into the set of knowledge that the University offers, it is recognized by other spaces and actors of the regional and international physics, it is a reference for the funding bodies of science existing at that time and expanding in the continent, such as the Rockefeller and Guggenheim foundations, UNESCO, or international companies like Westinghouse. Academic associations begin to consider the engineers who work there as physicists, including them in their rosters, inviting them to conferences and discussing with them on a level intellectual scientific ground.

But on the other hand, again, we see that this autonomy was relative to the conditions and material possibilities of development. As an example, even though there were initial interests in developing theoretical and nuclear physics, these purposes were set aside for much more experimental and less ambitious searches such as practical developments in nuclear medicine, crystallography and spectrography. Uruguay seems to have participated in international research, such as cosmic rays, but only nominally, as no records of work on cosmic ray detection in the country have been found (yet). This encourages us to think that what defined the physical sciences in the country had much more to do with the development of what was possible than those conceptual ideals that came from countries with more scientific development. But the possible was not only determined by the material. Another aspect deserves attention and it has to do with the national cultural environment and the intellectual environment of the University. It must always be kept in mind that, in a small country like Uruguay, the concentration of tertiary education in a single University, gave this a crucial role for the intellectual and professional development of the nation. The most pronounced characteristic of the University of the Republic, in a country characterized by always being on paths towards more development, has been its pedagogical character oriented towards the professional. With this tradition in tow, thinking of a university that could develop science in general, and basic science in particular, always generated tensions and disputes about the very functions of the country's largest house of studies.

The creation of the Faculty of Humanities and Sciences, in 1945, promoted and announced the development of those "disinterested" knowledge in the University. Given this professionalist tradition that we described, a new Faculty seemed to be the space where those knowledge, valuable in themselves, could come to be cultivated and developed. Therefore, with its creation, chairs of Physics-mathematics, mathematics, astronomy, chemistry or biology were included, which until then were proclaimed in relation to professional formations such as engineering, surveying, pharmaceutical or medicine. But the interesting thing about this proposal is that, for those years (end of World War II), the sciences could not be practiced in the way they were thought for the new Faculty. What happened with that institution was that, according to the humanistic and enlightened tradition of the University and the country, what flourished there were knowledge in letters, language, pedagogy, but little could be achieved in the branches of exact and natural sciences. The reason for this failure has to do with something substantive and characteristic of the time: if the new Faculty of Humanities and Sciences promoted, against the professionalist trend, "disinterested" studies, the few practitioners of the sciences in Uruguay never saw themselves as non-professional teachers or scientists. This tension was manifested not only in the heart of the new Faculty (as shown by the episode between Félix Cernuschi and Desiderio Papp), but in the whole of the debates about science that arose from the fifties. The scientific community claimed the possibility of professionalizing science, of making science a profession and moving it away from dilettante and vague models that had characterized the approaches in that house of studies.

The terrain of this dispute, between the scientific culture and the scholar, it was particularly given in philosophical discussions. To defend a professional science, a science that gives answers to phenomena and that pursues an adequate method of explanation, it was necessary to purify the philosophical vestiges that were hegemonic in the country. As we saw, this was precisely the task that some physicists like Loedel Palumbo or Félix Cernuschi embodied: to provide a philosophical foundation adequate for the public defense of science, a foundation that both found in the expressions of some logical empiricists, like Hans Reichenbach or Philipp Frank. This episode of the development of physics is not only remarkable because it was an effort of the community Latin American physicists to publicly defend science in the country, but because it shows the attempts to adopt and adapt European and North American philosophical programs to the context of the region.

Like in many parts of the world, the impact of the end of the war and the threat and subsequent launch of the atomic bomb, influenced societies about the dangers associated with nuclear development. In the Rio de la Plata region, this discussion reignited the possibility of discussing better and in a planned way how a national science from the Third World could be possible. Despite the continuous problems and obstacles, the community of physicists in Latin America, towards the sixties, was stronger. It was another type of impediments that attenuated its better development. The Cold War, the world divided between North American imperialism and Soviet socialism, seeped into the programmatic discussions of Uruguay and the region. If until that moment, part of the modest development of physics and basic sciences in general, had been achieved from financing from foundations like the Rockefeller and the Guggenheim, the ideological polarization and an organized student movement against North American imperialism, put the debates about the scientific autonomy of developing nations at risk. In this context, the University promoted the creation of the first Central Institutes of Basic Sciences, a project that did not materialize but that allows to know how, in those years, the main scientific actors of Uruguay and the region positioned themselves.

This compilation does not aim to summarize but, above all, to attend to a question that seems central to us: once the history of physics in Uruguay is constructed, in the way we have presented it in this book, are there criteria that allow us to affirm that, effectively, there was physics in the country? Is this a history of The Science? Does Uruguay stand on the shoulders of the Giant, or can it not even surpass the height of his ankles?

7.3 Who Is This *Giant*? Between the *Science as Results*, the *Science as Practice* or the *Science of the Possible*

Goethe's phrase, analyzed in depth, presents a double interpretation. On the one hand, when saying that "the history of science is science itself", we may simply mean that what we currently understand as science is only the result of those facts and episodes that historically lead us to the present definition of science. But a second interpretation could significantly change the meaning of the phrase, if we consider that any history that can be constructed can be taken as the way in which we should define science. More clearly, when the first interpretation adjusts the present definition of science to the past facts that led us to the current state, in the second it is intended to show that the current definition is only an arbitrary historiographical resource and that other narratives could lead us to a present definition of science, different from the one we have. We would like to refer to these two modalities as "science of results" and "science of practices", to show how each one has shaped the way in which we make history of science, but also to mark the need for a distinctive formulation to understand the science produced in Latin America, which escapes any of these two categories. In this last section we intend to reflect on the central question, for us, of all philosophy of science: what do we understand or should understand by science, in the Latin American context?

7.3.1 Science as Results

When Newton formulated that phrase in his letter to Robert Hook, he was probably wanting to say that this *Giant*, which is Science, is constituted by the set of findings of others like him, who managed to see beyond. Science is constituted from findings: which typically take the form of theories or facts. Theories and facts are knowledge, that's why this philosophical interpretation of science is usually defined with the label "science as knowledge".[1] The center of this notion of science lies in scientific theories, insofar as without them it seems to be very little that we could say about knowledge. In the introduction to his 1977 book -a classic in this perspective-, Fredrick Suppe defined the central role that theories play in understanding that which we conceive as science:

> If any problem in the philosophy of science can justifiably be claimed the most central or important, it is that of the nature or structure of scientific theories. For theories are the vehicle of scientific knowledge, and one way or another become involved in most aspects of the scientific enterprise. Were it not for theories there would be no problem of theoretical entities or terms, no problem of confirmation, no problem of cognitive significance. Models are employed scientifically either as a species of theories or as aids to the development of theories. Explanations centrally involve the subsumption of phenomena under theories. In

[1] Pickering (1992).

> the absence of theories to be tested or applied, experimental design is pointless. Laws and counterfactuals occur scientifically either as theories or as components of theories. It is only a slight exaggeration to claim that a philosophy of science is little more than an analysis of theories and their roles in the scientific enterprise. A philosophy of science's analysis of the structure of theories is thus its keystone; and should that analysis of theories prove inadequate, that inadequacy is likely to extend to its account of the remaining aspects of the scientific enterprise and the knowledge it provides. At the very least, it calls for a reassessment of its entire account of scientific knowledge.[2]

Suppe locates the philosophical tradition of science that held this type of thesis, neopositivism, in the cultural environment of the German academy. In his analysis of the historical background of neopositivism and the historical and cultural reasons that helped shape a philosophy centered on the analysis of theories, Suppe recognizes a series of foundational elements resulting from a varied number of factors that contribute to the birth of the idea of science as knowledge: the life and hierarchical structure (almost monastic) of German universities; the incipient developments that occur in the second half of the nineteenth century in some areas such as physiology and psychology; the fight against philosophical traditions always present -from Hegelianism to the new forms of neokantism-; in addition to the decisive advances in the field of physics. That is, Suppe makes a social history of the philosophy of science oriented towards theories, showing the course (not always univocal and clear) of a philosophy that clearly marked a new direction in contemporary thought seeking to align itself with the findings and revolutions that were taking place in various scientific areas. This new philosophy, scientific philosophy, always sought to give a place of relevance to theories, despite finding a huge number of expressions that at first glance seem distant from each other: authors like Mach, Reichenbach, Carnap, Neurath, Popper, Lakatos, Feyerabend or Hanson, with their undeniable differences and notable disagreements, are part of this German-Viennese cultural environment that sought to sustain the key to the scientific enterprise on the analysis of theories and their relationship with phenomena.[3]

If until 1935 the philosophical success of the new scientific philosophy, year in which its first meeting is held in Paris, was relatively tenuous, without a doubt the exile of a large number of its representatives to American universities served as an international promotion that until then had been denied to them by political and racial persecution in Europe.[4] The idea that science is explained by the knowledge that is contained in theories developed firmly on the basis -predominantly in physics- of the findings that managed to revolutionize several areas of the sciences.

---

[2] Suppe (1977).

[3] Under no circumstances do we mean by this that the appreciation of theories implies the unification of methods for philosophy to provide answers about them. Even less do we want to imply that the criteria of strictly verificationist significance to dismiss ethical, political or metaphysical theses was the path of the Berlin-Viennese scientific philosophy movement. The recent historical approach allows us to see the complexity and dissimilarity of the origin of neopositivism, as well as the biased publicity it suffered thanks to the reductionist and privileged version of English positivism. See: Stadler and Nielsen (2001), Reisch (2005), Verhaegh (2024).

[4] Verhaegh (2020).

Similarly, this same criterion was used to understand and risk theorizations in disciplinary fields much more labile than those of the hard sciences, such as the social and human sciences, or psychology. But the relevant thing about this narrative, for our purposes, is that scientific philosophy acted on consolidated or plausible knowledge to be so, that is, on findings made by scientific communities. This way of conceiving science was accompanied by a historiographical narrative that supported the narrative.[5] In the Latin American context, the first major work in Spanish, the *History of Physics* by Paul Schurmann, is without a doubt the monumental work of a physicist and historian of science in specifying in extremely precise terms who were and how were the great protagonists and contributions that helped to build that *Giant* on which to stand to see further beyond.

At the dawn of the twentieth century, the challenge was presented by the general theory of relativity. This theory, being conceived at the dawn of these Berlin and Viennese movements that lead to scientific philosophy, came to serve as an example of how all philosophy that prides itself on such should instrumentally operate on theories. As an example, the vast majority of the representative authors of logical empiricism did the philosophical work of analyzing the concepts and schemes of the Einsteinian theory long before building the philosophical framework that would characterize the movement: Moritz Schlick writes in 1915 *The philosophical significance of the theory of relativity*, and in 1917, one of his most outstanding works, *Space and time in contemporary physics*. Reichenbach writes in 1922 a work titled *The current state of the discussion on relativity,* and Carnap, a year later, *On the task of physics and the application of the principle of maximum simplicity*.[6] It is interesting to note that the treatment given in Uruguay to this new theory, in the same years, is far from assuming it as established knowledge, except for that young and small number of engineers who were interested in physics. The text by Geille Castro (published the same year as the one by Carnap), although it is a master class on the implications of the theory for the adventure of human knowledge, is austere regarding its significance due to its radically mathematized character. Geille Castro warns, from a pragmatic perspective, that every physical theory has an impossibility when noted mathematically, and that is that it can never be given rigorously. In other words, the "artificial" instrument of mathematics, for this physicist/engineer, transforms physics into a discipline at risk of not accounting for the world, of not being able to capture nature, in short, of not being able to act upon it.

7.3.2 Science as Practice and Culture

Many may have been the variables that led the philosophy of science to rethink the reduction of science to its theories. Undoubtedly, the most characteristic story of the history of philosophy tells us about the crisis that logical empiricism suffered in the

[5] Queijo Olano (2021b).

[6] Mormann and Peláez (2016).

fifties, now based in the United States, from the criticisms received by authors such as Willard Van Orman *Quine* and the work of the "second Wittgenstein". The philosophy of science attributes this crisis to the appearance, in 1962, of *The Structure of Scientific Revolutions*, by Kuhn. It is not the purpose to want to delve deeply into these historical reasons, but to note the appearance of new perspectives that try to define science not anymore from its theories, but from the practices carried out by scientists. This set of perspectives that conceived this definition has been called the proponents of the *science as practice (and culture).* It is perhaps the double axis of the Bath-Edinburgh schools, the epicenter from which a sociology of scientific knowledge is built in a structured way as a philosophical program of the sciences. Leaving aside the figure of Bruno Latour, who emerged in the seventies from his book *Laboratory life: The construction of scientific facts* (1979), this British school impacts studies on science to highlight the historically forgotten (or denied) features of science: its experimental character, its community and social character, and its inevitable cultural character.

By shifting the focus from theories to the spaces where science was developed, the instruments with which it was worked or the social, political and economic condition of scientists, the strong program of the *Sociology of Scientific Knowledge* (SSK) allowed to include also new approaches, such as that of authority or scientific legitimacy. Without a doubt, thinking about science from its practice enabled new narratives and a whole historiographical development that accompanied this new understanding that science are scientific practices. It is unnecessary to account for all that academic production that developed from this new way of understanding science, as well as the whole series of ramifications of academic programs that derived from this *strong* programmatic impulse of the seventies.[7] We want to highlight an aspect that can be adequately inferred from the analysis of some of the canonical works of the strong program of the *SSK*.

Steven Shapin and Simon Schaffer have been one of the most committed representatives of the program, and their book *Leviathan and the Air - Pump*[8] has been an unavoidable reference for those who have intended to work in this approach. This historical work accounts for the disputes over authority and scientific legitimacy in the discovery of the vacuum pump, which involved Robert Boyle and Thomas Hobbes in seventeenth-century England. Not only that, the work delves deeply into a central scientific issue, namely, how a scientific truth is constructed from an experiment, under what conditions another scientist accepts or rejects the experimental methods that allow the definition of that truth. On this exercise of the history of science, it is intended to establish the series of premises that guide the work of the strong program of the SSK. The experiments and methods involved in a scientific practice lead to truths, only if we understand that they do not emerge evidently for the community or the actors involved. Other elements, historically forgotten by a narrative of science as results, come into play relevant to construct truth, legitimacy

[7] An exhaustive and relatively brief summary of the production of this sociological approach to scientific knowledge can be found in Shapin (1982).

[8] Shapin and Schaffer (1985).

and scientific authority. The social position, cultural prestige, recognition of third parties, and the way of conducting experiments are part of those elements that underlie the facts that we consider scientific milestones. Shapin and Schaffer, in this work, effectively account for that background, managing to show that behind the results there are cultural practices that naturally mix with scientific work.

Of equal impact was the book by Harry Collins, *Changing order: Replication and induction in scientific practice*, from 1985. It does not function so much as material of the history of science, although his work must make use of a historical style. His work as a sociologist of science has focused on the historical process that has led the community of physicists to the detection of gravitational waves. In this years-long work, Collins has noted that the physicists involved in this search generally reproduce experiments and infer premises from the data obtained in a very random way, which only manages to come together when community-type agreements are reached, where the data seem to have been left behind. Collins' approach, much more oriented to the modern physics of the twentieth century, aims to show the eminently community character of practices such as replication or induction in scientific activity. Without that community character, the truths in the field of gravitational waves could not be possible in any way.

We would like to step a little out of the strong program of the SSK, to highlight another central work of the period as it was *Representing and intervening: Introductory topics in the philosophy of natural science*, by Ian Hacking.[9] In this work, now more emblematic of the philosophy of science than from a sociological perspective, Hacking discusses in the first part the problem of scientific representation -the theories, science as results-, to then propose a central dimension of scientific activity, its intervention in nature -science as experiment, as practice-. Thus, this work takes us from the discussion on how philosophers have discussed the forms of what scientific theories represent, the way of understanding them and the various ways of understanding their link with what we call reality; to move in the second part to show us that the philosophical discussion for realism is controlled if we understand the eminently experimental character of science and the intervening power in natural phenomena. Hacking, without being in his central purposes, elaborates in this second part of the book a recap of episodes of science where he accounts for how, effectively, those great thinkers of scientific activity were above all great practitioners of science: Bacon, Leibniz, Boyle, Hooke or more contemporarily Robert Millikan. Among all these authors, what Hacking is interested in highlighting is the practical character of their activity, it is how from not philosophical reflections but from scientific interventions they manage to conclude truths about the existence of entities or causal explanations that until that moment were hypothetical. Science intervenes in the world and in that practice we can recognize that the things with which we play to doubt philosophically, exist because they respond to us in some way.

[9] Hacking (1983).

With these three canonical authors, different from each other (much in the case of Hacking), we only want to highlight the changing aspect of what is understood by science since the seventies. Science, as we said, begins to be conceived as a practice, and this view of the science as practice manages to penetrate firmly into the tradition of the philosophy of science.

The history of Latin American science, except for counted and heroic exceptions, has not been characterized by being a *science of results*. The history of physics in Uruguay in particular, and of the region to a large extent, does not show results that are commonly incorporated into the general books of the history of the discipline. But it is clear that this should not lead to the conclusion that in these countries no science was done. Perhaps, and only perhaps, could it be said that what existed, more precisely, were not *scientific results* but *scientific practices*? But, what is a scientific practice? Is the definition of *science as practice* adequate to understand what the development of science in Uruguay and in the rest of Latin America implied?

In the text of Andrew Pickering, cited above, there is a footnote where the author problematizes the concepts of "practice" and "culture". In logical empiricism, Pickering suggests,

> (...) scientific culture consists in a field of knowledge and knowledge claims, and scientific practice consists in the appraisal of conceptual knowledge claims against observational knowledge, an appraisal ideally governed by some logic or method.[10]

While this characterization can be discussed, and in fact has been by several recent studies on the movement of scientific philosophy and that we have already indicated above, it allows us to arrive at the issues that interest us. The view of the sociology of scientific knowledge, built in part from the evocation of an enemy -logical empiricism- characterized in a very reductionist and arbitrary way, sought to provide new views on these notions of "practice" and "scientific culture". It is precisely in this aspect that constructivism in science sought to mark its difference with the logical empiricist philosophical movement. Pickering indicates that for this constructivism, the redefinition of these concepts is determinant

> (...) Centrally at issue is the constructivist insight that doing science is real work and that real work requires resources for its accomplishment. Throughout this essay, "culture" denotes the field of resources that scientists draw upon in their work, and "practice" refers to the acts of making (and unmaking) that they perform in that field. "Practice" thus has a temporal aspect that "culture" lacks, and the two terms should not be understood as synonyms for one another: a hammer, nails, and some planks of wood are not the same as the act of building a dog kennel-though a completed dog kennel might well function as a resource for future practice (training a dog, say). (...) It might also be useful to emphasize that my usage of "culture" here is a deflationary one. It encompasses all of the resources, many of them humble and mundane, that scientists deploy and transform in their practice (...). It is not a way of gesturing at grand, all-encompassing worldviews, for example, or at big cultural currents that flow between science and the outside world-though neither is it a way of denying that unifying characterizations of entire cultures might also be perspicuous on occasion.[11]

[10] Pickering (1992, 3).

[11] Pickering (1992, 2–3).

Science as practice and culture then denotes the set of material and conceptual resources that a scientist has (this would be his scientific culture), while allowing him to develop his activity (this would be his scientific practice). For the purposes of our interest, determining with which conception of science we should work to understand the history of physics and science in Latin America, the constructivist definitions only seem to be apparently more adequate. Although in them neither scientific knowledge, nor scientific statements, nor anything Regarding theories, we are interested in highlighting that this perspective originating from the strong program in SSK (and the historical studies that support it) does not mention *the results of science* but presupposes them in its definition of *science as practice and culture*. Although this school emphasizes the resources and ways of doing of scientists, it forgets that these resources and ways of doing lead to results. In fact, in the canonical examples of this perspective, all historical studies are about cases and characters who have obtained results. This view of science as a builder of knowledge does not necessarily disappear or is surpassed by giving greater emphasis to material resources and scientific ways of doing. Shapin, Schaffer, Collins, Hacking or Pickering, rescue the practical character of science through history, but work on names or discoveries that, ultimately, constitute the great results that shaped the *Giant*: laws, theories, models, instruments, etc.... More categorically: both the *science of results* and the *science as practice and culture*, are nothing more than a history of scientific success.

7.3.3 Science of the Possible

If, ultimately, the historiography of science -with various approaches and accents-, will continue to highlight only the successes scientific as milestones in the general history of science, the history and scientific reality of Latin America will be destined, luckily, to occupy a footnote in that great book of successes. What we would like to recognize, instead, is that a much more complex view is necessary to understand science on the Latin American continent. Modestly, and as a way to foster future debates, we will call this perspective: *science of the possible*.

What would be the best way to define *the possible* to understand in a general way that set of activities that allow integrating the history of scientific development in countries like Uruguay, in a field like physics?

A first approach can be given by what Peter Medawar—biologist and immunology specialist, born in Brazil and later based in England, awarded in 1960 with the Nobel Prize in Medicine for his contributions to skin transplants-, has characterized *the art of the soluble* (1967). Medawar, evoking and criticizing Arthur Koestler's work, *The act of creation* (1964), presents a definition of scientific activity in the following terms:

> No scientist is admired for failing in the attempt to solve problems that lie beyond his competence. The most he can hope for is the kindly contempt earned by the Utopian politician.

> If politics is the art of the possible, research is surely the art of the soluble. Both are immensely practical-minded affairs.[12]

This definition allows us to introduce two elements that will help us to characterize our idea of *the possible*. The first element is the contempt and lack of recognition for failed attempts, or more generally, for scientific practices that do not achieve recognition in the international scientific narrative. The history of physics that we have traced is full of episodes of contempt and disdain towards the physicists who tried to conform the discipline in Uruguay. To the extent that their activity did not meet the expectations of the "utopian politician", the situation of physics in Uruguay was always subjected and conditioned in its possibilities. The second element to highlight in this phrase from Medawar is the one referred to thinking about research as the "art of the soluble". Research in physics in Uruguay, we can say, was characterized by focusing on the soluble, what could be effectively carried out with the means available for it.

The science of the possible will be, therefore, that set of activities oriented not so much to the factual and hypothetical possibilities that mark the destiny of a discipline, but to the material anchoring that conditions and marks the adaptation of research to the resources (human, material, institutional and cultural) available. From the perspective of the researcher, the scientist, this terminological nuance is totally determinant. It is determinant for a Uruguayan physicist to understand that there were null or almost null possibilities to develop theoretical and nuclear physics at the University of the Republic in 1940. It is determinant for a physicist to adhere to an international research network in cosmic rays not to produce results, but to obtain some equipment from the funding that can be obtained. It is also determinant to start lines in areas like nuclear medicine or spectrography if they are areas that can be financed from abroad, and can be developed with the available material. But it is also determinant to defend national science, inside and outside the University, to sustain the scarce possibilities that had been conquered so laboriously.

This, we believe, is a definition that fits the historical contexts that we have tried to outline in this book. *The science of the possible* varies from context to context, from discipline to discipline, it is rooted not only in resources, but in an *ethos* that is only defended by scientists.[13] It is natural that the European scientific tradition has conceived and firmly embraced the figure of the *Giant*, because it sees as *possible* towards the future a construction on its shoulders. Moreover, that *Giant* not only allows a projected look into the future, of long breath, but it also allows looking into the past with the same depth. European science imagines itself as a *Giant* that allows to see "beyond" as long as there is a past and an ethos that is endorsed and socially legitimized. The future of any scientific development is closely linked with tradition, that is the fundamental teaching of Goethe.

In contexts without tradition, where history is a narrative imported from elsewhere, science becomes an empty and even disconcerting image. That image of the

---

[12] Medawar (2021, 101).

[13] Queijo Olano and Videira (2022b).

Giant is not *possible* nor does it work outside the context where it was conceived, it does not work for the case of physics in Uruguay, no matter how international Science is intended to be conceived. Even more, that image of the *Giant* operates as a burden for those scientists far from that tradition, who seek to do from the *possibilities* materials they have at their disposal. In addition to trying to make their relevance understood, these scientists must build a *scientific ethos* in an environment that is deaf to any demand of science. When science is structured from *the achievable*, from the *immediate*, there are not many possibilities to be able to look "beyond". At most you can see in the immediate future, in the longing not to be able to build on a day-to-day basis.

And that great *Giant,* which had been conceived as a positive image for the scientific enterprise, ends up crushing, standing on the shoulders of those who will never be able to see beyond.

Bibliography

Hacking, I. 1983. *Representing and Intervening: Introductory Topics in the Philosophy of Natural Science*. Cambridge: Cambridge University Press.

Medawar, P. B. 2021. *The Art of the Soluble*. New York: Routledge.

Mormann, T. H., and Á. Peláez. 2016. *Logical Empiricism. Basic Texts*. Bogotá: Editorial Universidad del Rosario.

Pickering, A. 1992. From science as knowledge to science as practice. In *Science as Practice and Culture*, ed. A. Pickering. Chicago & London: The University of Chicago Press.

Queijo Olano, J. A. 2021a. Cernuschi vs. Papp: The Uruguayan Dispute over the History of Science. *Transversal: International Journal for the Historiography of Science* 11.

Queijo Olano, J. A. 2021b. The history, the science and its shades. In *Contemporary Perspectives in Philosophy of Science: Volume II*, ed. M. C. M. Motta, A. A. P. Videira, and F. Fragozo, 1st ed. São Paulo: Livraria da Physics.

Queijo Olano, J. A., and A. A. P. Videira. 2022a. The ethics of scientific practice: Enrique Gaviola and physics in Argentina between 1930 and 1956. *Knowledge. Journal of the History of Sciences and Humanities 5* (12): 89–113.

Queijo Olano, J. A., and A. A. P. Videira. 2022b. Walter Scott Hill and Uruguayan Physics. *Physics in Perspective* 24 (1): 35–71.

Reisch, G. A. 2005. *How the Cold War Transformed Philosophy of Science*, 268. Cambridge: Cambridge University Press.

Shapin, S. 1982. History of science and its sociological reconstructions. *History of Science* 20:157–211.

Shapin, S., and S. Schaffer. 1985. *Leviathan and the Air-Pump: Hobbes, Boyle, and the Experimental Life*. Princeton, NJ: Princeton University Press.

Stadler, F., and C. R. Nielsen. 2001. *The Vienna Circle: Studies in the Origins, Development, and Influence of Logical Empiricism*, 279–299. Vienna: Springer.

Suppe, F., ed. 1977. *The Structure of Scientific Theories*. Vol. 634. University of Illinois Press.

Verhaegh, S. 2020. Coming to America: Carnap, Reichenbach and the Great Intellectual Migration. Part I: Rudolf Carnap and Part II: Hans Reichenbach. *Journal for the History of Analytical Philosophy* 8:11.

Verhaegh, S. 2024. Logical Positivism: The History of a "Caricature". *Isis 115* (1): 46–64.

Name Index

J. A. Queijo Olano, E. Mozo Meneses, *Giants Upon Our Shoulders: A Philosophical History of Physics in Uruguay*, Studies in History and Philosophy of Science 65, https://doi.org/10.1007/978-3-031-97729-9

Subject Index

J. A. Queijo Olano, E. Mozo Meneses, *Giants Upon Our Shoulders: A Philosophical History of Physics in Uruguay*, Studies in History and Philosophy of Science 65, https://doi.org/10.1007/978-3-031-97729-9

The manufacturer's authorised representative in the EU is Springer
Nature Customer Service Centre GmbH, Europaplatz 3, 69115 Heidelberg,
Germany. If you have any concerns regarding our products, please
contact ProductSafety@springernature.com

Printed and bound by CPI Group (UK) Ltd, Croydon, CR0 4YY
01/10/2026
02225865-0002